Progress in
Colloid & Polymer Science

Editors: H.-G. Kilian (Ulm) and A. Weiss (Munich)

Polymers as Colloid Systems

32nd Meeting of the Kolloid-Gesellschaft and
the Berliner Polymeren Tage,
Berlin, October 2–4, 1985

Guest Editor: J. Springer (Berlin)

Springer-Verlag Berlin Heidelberg GmbH

ISBN 978-3-662-15201-0   ISBN 978-3-7985-1695-3 (eBook)
DOI 10.1007/978-3-7985-1695-3
ISSN 0340-255 X

© 1985 by Springer-Verlag Berlin Heidelberg
Originally published by Dr. Dietrich Steinkopff Verlag GmbH & Co. KG, Darmstadt in 1985
Softcover reprint of the hardcover 1st edition 1985
Copy editing: Cynthia Feast; Production: H. Frey

Contents

Progress in Colloid & Polymer Science   Progr Colloid & Polymer Sci 72:1–11 (1986)

Transfer of the nucleic acid information into protein structure, and some aspects of the structure and function of the translating organelle*)

K. H. Nierhaus

Max-Planck-Institut für Molekulare Genetik, Abteilung Wittmann, Berlin

Abstract: The metabolism of a cell is characterized by two features: 1. The network of chemical reactions is defined by the pattern of enzymes. 2. In vivo the biochemical reactions run far from their thermodynamical equilibrium, which is made possible by the prevalence of the ATP over its hydrolytic products. If we compare the metabolism with a channel network, then the enzyme pattern defines the architecture, and the ATP to ADP ratio the flow rate of the system. Practically all the enzymes are proteins. Proteins have additional regulatory functions and important structural tasks, they comprise 50 % of the cell's dry mass. For these reasons, in molecular biology heredity means first and above all the heredity of structural information on both proteins and the protein-synthesizing machinery. The fundamental molecular processes of heredity, i. e. the preservation and the realization of the genetic information, are based on the simple rule of complementary base-pairing, resulting in the translation of the genetic information into the defined amino acid sequence of a protein. The amino acid sequence in turn determines the routes of folding and thus the spatial structure of a protein ("folding code").

The ribosome is the organelle of translation. Three aspects are surveyed: 1. Principles of assembly of the large subunit from procaryotic ribosomes (*Escherichia coli*). 2. Neutron scattering as a method for the analysis of the internal topography of ribosomes. 3. Description of ribosomal functions in the frame of a model containing three tRNA binding sites.

Key words: Translation, ribosomes, nucleic acid, protein structure.

1. Introduction

Proteins represent more than 50 % of the dry mass of all cells. The importance of this class of molecules becomes particularly evident when one recalls that enzymes are almost exclusively proteins. Enzymes are biocatalysts, i. e. they do not influence a reaction equilibrium, but rather accelerate the reaction rate by lowering the activation energy.

The reduction of the activation energy is carefully directed. If, for example, a substrate X can enter four different pathways each with a high activation energy, an enzyme will specifically diminish the activation barrier of only one pathway, with the result that substrate X will react exclusively to the product Y. The specific acceleration of the reaction rate is extremely large and can reach a factor of 10^6 to 10^{10}. This enormous effect is the reason why the metabolism, which is at first glance a chaos of reactions, can occur efficiently in the cell. Up to 1500 reactions run simultaneously in a bacterial cell (*Escherichia coli*), whereas there are up to 10000 reactions in a human cell. Almost all of these reactions are steered by a defined enzyme.

From the foregoing statements we can derive the first necessary, but not yet sufficient, condition for a formal description of the metabolism: the enzyme pattern defines the cell's metabolism. If we compare the flux of matter in the metabolism with a channel network, where the crossing points represent the chemical reactions, then the enzyme pattern fixes the architecture of the channel network. This concept was termed "the axiom of the biochemistry".

*) Lecture presented during the 32nd Annual Meeting of the Kolloid-Gesellschaft, Berlin October 2-4, 1985.

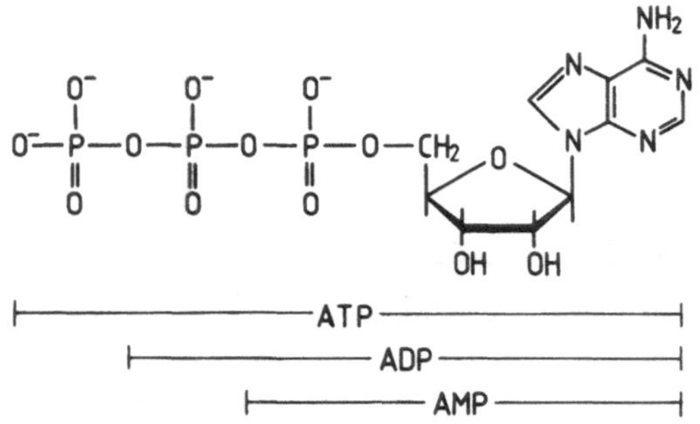

Fig. 1. Structure of adenosinemono— (AMP), adenosinedi- (ADP) and adenosinetriphosphate (ATP)

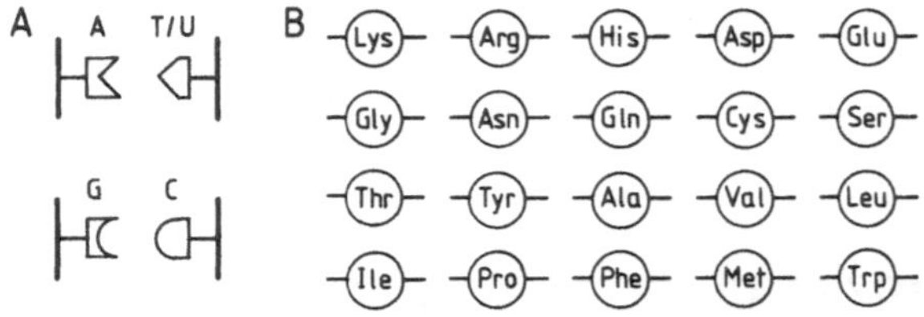

Fig. 2. Nucleic acids consists of four different (A) and proteins of 20 different components (B)

A second essential feature of the metabolism is seen in the fact that all reactions within the cell are running far from their equilibrium. Reaching equilibrium would result in cell death. Therefore, the maintenance of the chemical non-equilibrium is a vital task of the metabolism and requires a continuous energy supply.

The central start- and end-point of the energy turnover is the energy-rich molecule ATP (adenosinetriphosphate, Fig. 1). The two acid anhydride bonds between the three phosphate groups are surrounded by negatively charged, repulsive oxygen atoms. Consequently, these two acid anhydride-bonds of the ATP are labile, i. e. they are energy-rich. The useful energy yielded upon hydrolysis amounts to $\Delta G = -11$ to -13 kcal/mol for each of the acid anhydride bonds. This value already takes into account the cellular concentration of ATP and its products. The processes generating energy-rich compounds (ATP) are of central importance for the metabolism; these processes are photosynthesis, anaerobic glycolysis and respiration.

This leads us to the second necessary condition, which, together with the first one, allows a sufficient, formal description of the metabolism: a cellular ratio of ATP to ADP of 8–10 to 1 guarantees the maintenance of the non-equilibria of the reactions. This ATP/ADP ratio resembles a battery which drives the life motor. Just as the enzyme pattern defines the architecture of the channel network, so does the ATP/ADP ratio determine the flow rate in the channel system.

2. The principle of heredity

The eminent importance of the proteins outlined above for the events in the cell is reflected in the fact that the heredity processes can be reduced essentially to the heredity of protein structure. In fact, in molecular biology heredity means first and above all the heredity of structural information in both proteins and the protein synthesizing apparatus. This implies an essential reduction of the heredity processes to two classes of molecules: the *nucleic acids*, for information storage and messenger function, and the *proteins* as realization of this information.

Both nucleic acids and proteins are long, non-branched molecular chains. Nucleic acids consist of only 4 different building blocks, the nucleotides (Fig. 2A). The sequence of the nucleotides contains the information for the protein structure. In contrast, we find 20 different building units (amino acids) in the proteins (Fig. 2B). Thus, the problem is to translate a language of four letters (nucleotides) into one of 20 letters (amino acids).

The building units of the nucleic acids are characterized by the notable capability to combine in a pairwise fashion. G (Guanine) and C (Cytidine), and also A (Adenine) and T (Thymine; U Uridine), form base pairs. The reason is that in these combinations the base pairs most effectively form hydrogen bonds. One says that G and C and also A and T (U) are complementary (Fig. 2A).

Two closely related species of nucleic acids, the DNA (deoxyribonucleic acid) and the RNA (ribonucleic acid), participate in the storage and transfer of information. In 1958 Crick postulated a connection of DNA, RNA and protein synthesis in the "central dogma" of genetics: "DNA makes RNA, RNA makes protein". This dogma was substantiated in the early sixties.

DNA stores the cellular information for the heredity processes. The DNA consists of the four nucleotides G, C, A and T and contains the sugar deoxyribose.

The bacterium *Escherichia coili,* which is the model organism of the molecular biology, contains two DNA molecules each with 3 millions of nucleotides.

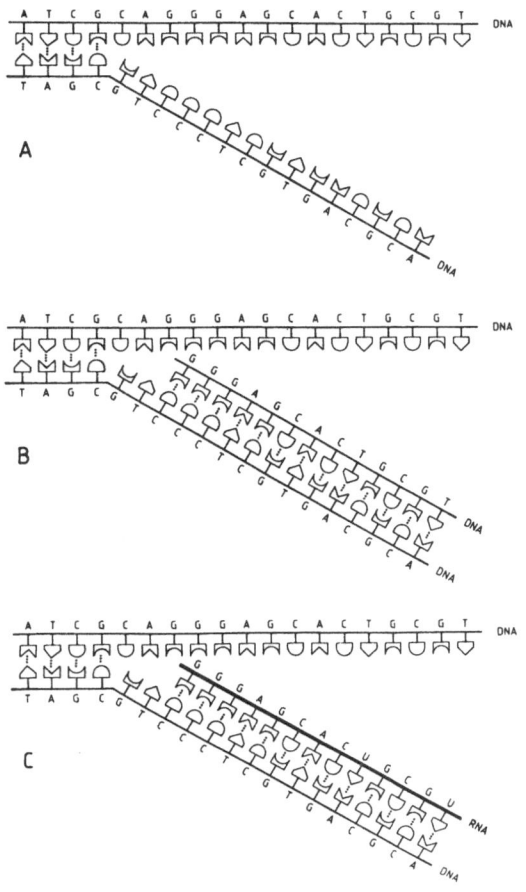

Fig. 3. A, The DNA (double helix) opens for a copy process. B, replication: both strands are copied. C, transcription: one strand is copied

The two strands are organized in a double helix, which is closed to form a ring. The essential point is that one DNA chain is complementary to the other, i. e. an A in one chain is opposed to a T in the other and a G in one chain to a C in the other (see Fig. 3A). Therefore both chains contain indentical information, they are analogous to the positive and negative of a photograph.

When a cell divides, the information in the DNA is transferred to the two daughter cells in the following way: The double helix opens like a zipper and each old chain is taken as template for the synthesis of a new strand according to the rules of complementary base pairing (replication). Both daughter cells receive a double helix consisting of an old and a new chain (semi-conservative replication, see Fig. 3B). In this way the life information has been transferred from one cell to another for the last billion years.

When the DNA information is to be converted into protein structure, the first step is a process similar to the replication. The DNA double-helix opens, but now only one chain is used as a template to direct the synthesis of a complementary RNA molecule (transcription, see Fig. 3C). The structure of the RNA is the same as that of the DNA, except that RNA contains the sugar ribose and the base U instead of T. The resulting RNA molecule serves as a "messenger" (messenger RNA or mRNA). It hands over the information from the DNA to the protein synthesizing machines, the ribosomes.

The reason for the change of the nucleic acid species from DNA to RNA in the course of transcription resides in the different stability of the two molecular species. RNA is very labile and hydrolyzes easily at alkaline pH, whereas in the case of DNA the lack of the 2'—OH group (deoxyribose) leads to stability and resistance to alkaline hydrolysis. It is self-evident that an information storage system needs maximal stability. In contrast, a short half-life is advantageous for the information-transferring molecule (mRNA), in order to be able to make rapid changes in the program of the protein synthesizing machinery in response to changing conditions. The half-life time of a bacterial mRNA lies in the range of two minutes, whereas in animal cells the half time is longer, and can be as much as a hundred days (e. g. the mRNA of globin which is needed for the synthesis of hemoglobin in red blood cells).

The ribosomes, where the proteins are synthesized, are the intercept of two fluxes. The first of these is the mRNA bearing the information for the protein structure. The second is represented by another kind of RNA molecule, which serves to effect the transfer of the amino acids from the cell sap to the ribosome. This second species of RNA molecules is termed transfer RNA or tRNA, and at least one specific tRNA species exists for each of the twenty amino acids (Fig. 4).

A glance at Figure 2 tells us at once that in the course of translation one nucleotide cannot correspond to one amino acid, since four building units on the one side (in the nucleic acid) are opposed to twenty on the other (in the protein). Rather, a sequence of three nucleotides, a triplet, corresponds unequivocally to one amino acid. The triplet (or "codon") UUU, for example, specifies for the amino acid Phe (Phenylalanine). For this codon there exists one tRNA species, which has an L-formed structure (like all tRNAs) and which in this case has a Phe covalently bound at the short arm of the "L", whereas at the top of the long arm three nucleotides are exposed. These three nucleotides are complementary

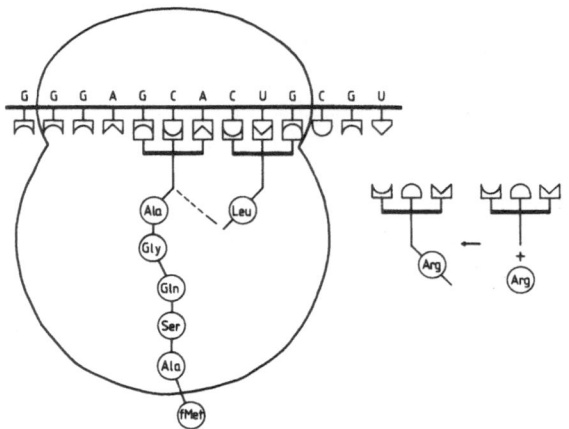

Fig. 4. Translation: Outside the ribosome a specific tRNA is linked to the corresponding aminoacid. The ribosome selects the appropriate aminoacyl-tRNA. The peptide chain is transferred to the new aminoacyl residue

1. Position (5' Ende) ↓	2. Position				3. Position (3' Ende) ↓
	U	**C**	**A**	**G**	
U	Phe Phe Leu Leu	Ser Ser Ser Ser	Tyr Tyr STOP STOP	Cys Cys STOP Trp	U C A G
C	Leu Leu Leu Leu	Pro Pro Pro Pro	His His Gln Gln	Arg Arg Arg Arg	U C A G
A	Ile Ile Ile Met	Thr Thr Thr Thr	Asn Asn Lys Lys	Ser Ser Arg Arg	U C A G
G	Val Val Val Val	Ala Ala Ala Ala	Asp Asp Glu Glu	Gly Gly Gly Gly	U C A G

Fig. 5. The codon-lexicon

to the codon UUU, and therefore are termed "anticodon".

The coding properties of all possible codons is compiled in the so-called codon lexicon (Fig. 5). Up to six different codons can be assigned to one and the same amino acid (degeneracy of the genetic code). For three codons (for example UAA) no amino acid exist. These three codons represent stop codons and signal the ribosome to stop the synthesis of a protein.

Figure 4 demonstrates the process of protein synthesis on the ribosome. The ribosome carries the tRNA to which the polypeptide already synthesized is attached. After the adjacent codon has bound the corresponding aminoacyl-tRNA, here Leu-tRNA, the ribosome transfers the already synthesized polypeptide to the new aminoacyl-tRNA, resulting in a polypeptide prolonged by one amino acid. Afterwards the ribosome moves along the messenger RNA by one codon length and binds the next aminoacyl tRNA complementary to the new codon. In this way the ribosome recognizes the information on the messenger RNA in codon units by means of the corresponding tRNAs, and converts the information into the amino acid sequence of a protein.

The essential result of this cursory survey is that the three fundamental processes of molecular heredity, namely replication, transcription and translation, all make use of the simple rule of complementarity between nucleic bases.

The genetic information has condensed in the sequence of the amino acids, i. e. the primary sequence of proteins. However, this is not the end of the information flux. The synthesized protein must fold to a defined, compact, active three-dimensional structure before it can take over its physiological role. The theoretical structure possibilities for a protein containing three hundred amino acids are so immense that a hundred years would not be long enough to probe all the possibilities even if one hundred different possibilities could be checked every second. It follows that only a drastically reduced number of folding possibilities can exist for one protein. The information for the folding process leading to an extremely dense, defined protein structure is determined in the sequence of the amino acids. This information content has been termed as the "second code", the folding code, as opposed to the genetic code. Today we are still far from an understanding of the folding code, and we are not able to predict the three-dimensional structure of a protein from its primary sequence. Some features of the folding code, however, have emerged and are summarized in Figure 6.

Two features are essential:

1. The folding process passes through a hierarchy of structures. At the beginning folding nuclei are formed in 10^{-3} seconds as a result of interactions between neighbouring amino acids. A typical example is the so-called "secondary structure" α-helix. Thereafter structures appear which take advantage of interactions between aminoacyl residues which are more distant. Examples of such structures would be the secondary structure β-sheet or the super-secondary structures that consist of defined combinations of α-helices and β-sheets. The folding of these structures needs up to

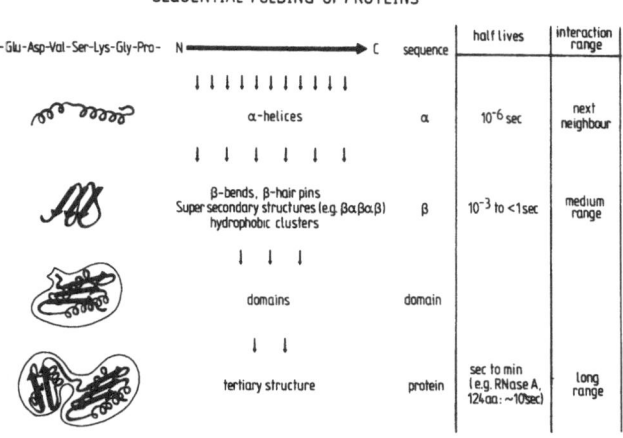

Fig. 6. Folding of a protein: hierarchy of the intermediate structures

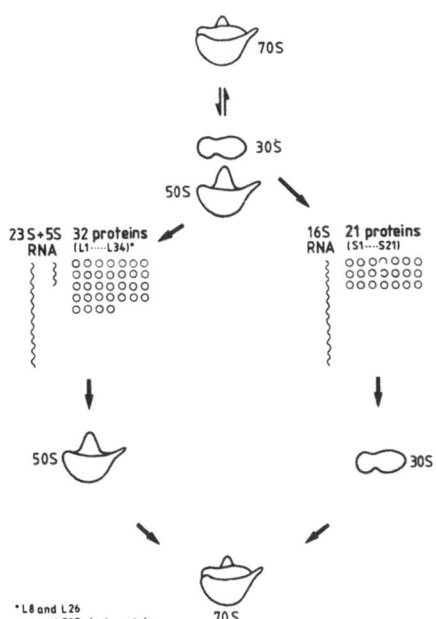

Fig. 7. Components of a ribosome from the bacterium *Escherichia coli*

one second to take place. Thereafter domains are formed which comprise at least 50 amino acids, and which represent compact building blocks within the mature protein. The domains associate to give the final 3D-structure of the protein. Often the active centres of enzymes are located at the interfaces of adjacent domains.

2. In contrast to some descriptions there is not a single defined folding pathway. Rather, there exists a small number of independently folding nuclei. The subsequent folding ways are confluent, ending in one defined 3D-structure. The total folding process for the protein RNase A (124 amino acids) for example, lasts for about 10 seconds.

3. The ribosome

The ribosome is one of the most complicated particles known in molecular biology, both in terms of its structure and function. More than 20 different functions of the ribosome can be tested in vitro. In the following sections I shall briefly survey some aspects of the assembly, structure, and function of the ribosome which have been analyzed in my group.

3.1 The assembly of the ribosome

As with all ribosomes the *E. coli* separates in two subunits, a small and a large one, twice as heavy. Both subunits are complexes of proteins and ribosomal RNA (rRNA). The small subunit consists of one rRNA molecule and 21 different proteins. These pro-

teins are numbered essentially according to their molecular weight and are denoted by the prefix S (from *s*mall subunit). S1 is the largest and S21 the smallest protein. The large ribosomal subunit is similarly constructed. Here we find two ribosomal RNAs and 32 different proteins, which are also numbered and carry the prefix L (from *l*arge subunit; see Fig. 7). All of these numerous ribosomal components are present in one copy per ribosome (with only one exception). In spite of the large number of proteins this class of molecules constitutes only one third of the ribosomal mass, the remaining two thirds being composed of rRNA.

It is of great importance that we can reconstruct functionally active subunits from the isolated ribosomal components in vitro. The method for such a total reconstitution of the small subunit was introduced 20 years ago, the corresponding method for the total reconstitution of the large subunit of *E. coli* ribosomes about 10 years ago. The total reconstitution leads to particles which are indistinguishable from native subunits with respect to structure and function. The principles of the ribosomal assembly could be unravelled by means of this technique.

The sequence of assembly was elucidated for both subunits, component by component ("assembly map"), and further important results from assembly studies of the large ribosomal subunit are compiled in

Table 1. Features of the 50 S Assembly

1. In vitro: 3 intermediates	$RI_{50}(1) \rightarrow RI_{50}^*(1) \rightarrow RI_{50}(2) \rightarrow 50\,S$
	(33 S) (42 S) (48 S)
In vivo: 3 precursors	$p_1 50\,S \rightarrow p_2 50\,S \rightarrow p_3 50\,S \rightarrow 50\,S$
	(32 S) (43 S) (~ 50 S)

2. The assembly starts with only two proteins (assembly initiator proteins):

    L24 and L3

3. The early assembly reactions depend on only five proteins:

    L4, L13, L20, L22 and L24
    (early assembly proteins).
    L3 stimulates.

4. The five early assembly proteins bind exclusively to the 5'end of the 23SrRNA:

    Assembly Gradient.

5. At least two of the early assembly proteins, L20 and L24, are mere assembly proteins. They are not involved in both late assembly and function of the 50 S subunit.

Table 1. The in vitro assembly process passes through three distinct intermediate particles, in accordance with the corresponding processes in the cell. Two proteins initiate the assembly, both of which must simultaneously be present on one 23S rRNA molecule for a successful start of the assembly. These two proteins, together with four further proteins, define the subsequent early assembly phase. Interestingly, five early-assembly proteins bind to one end of the ribosomal RNA, namely that end where the rRNA synthesis begins. Thus, the ribosomal assembly follows the synthesis of the ribosomal RNA, i. e. the association of the ribosomal components already begins before the synthesis of the ribosomal RNA is completed. This relationship was termed "assembly gradient", indicating that the progress of the rRNA growth dictates the progress of the ribosomal assembly.

The assembly gradient has an important consequence. In the cell the early assembly reactions start soon after the onset of the rRNA synthesis using the relatively short piece of nascent rRNA and the five assembly proteins, whereas in vitro the assembly process has to operate with the complete mature rRNA molecules and 32 different proteins. It follows that the entropic situation in the cell is much simpler than that in the test tube. This is obviously one important reason why the ribosomal synthesis occurs within two minutes at 37 °C in the cell, whereas we need 1.5 hours at 50 °C in vitro.

The assembly gradient, i. e. the parallel occurrence of rRNA synthesis and ribosomal assembly, is of such importance that it had to be maintained in the eucaryotic cell where the DNA is concentrated in the nuclei. Here the maintenance of the assembly gradient has led to a complicated assembly mode. The rRNA genes are present in special regions of the nuclei, the nucleoli, in contrast to the genes of the ribosomal proteins. Thus, the mRNAs for the ribosomal proteins have to be exported out of the nuclei into the cytoplasm, where the proteins are synthesized. Then the ribosomal proteins have to be imported in the nucleoli, where the coupled processes of rRNA synthesis and ribosomal assembly take place. The mature (or quasimature) ribosomal subunits are then exported to the cytoplasm, where they perform protein synthesis.

By means of the total reconstitution of the large ribosomal subunit it was possible to identify five components which are essential for the assembly of the peptidyltransferase centre. The peptidyltransferase centre is located on the large ribosomal subunit and exerts the central enzymatic function of the ribosome, namely the formation of the peptide bond between the amino acid residues.

3.2 Structure of the ribosomes

The prevailing method for the study of the outer topography of the ribosome is electron microscopy. With this technique the shape of both subunits has been determined (largest diameter about 250 Å). Furthermore, using specific antibodies against single ribosomal proteins, the surface location of the corresponding antigenic determinants (epitopes) could be pinpointed. The present state is presented in Figure 8. The upper row shows various views of the small and the lower row those of the large subunit. The numbers indicate the location of protein epitopes; the number 4, for example, in the pictures of the upper row marks the surface location of the antigenic determinant of protein S4 on the small subunit.

However, one should bear in mind that the biological samples for electron microscopy have to be subjected to harsh preparation conditions. First the samples are dehydrated, then they are soaked in uranyl acetate (which is a poison for the ribosomes), or alternatively shadowed with metal. Therefore one may not assume *per se* that the structures shown in Figure 8 are identical to those of ribosomes engaged in protein synthesis. It follows that it is desirable to complement our knowledge with information on the inner topography of the

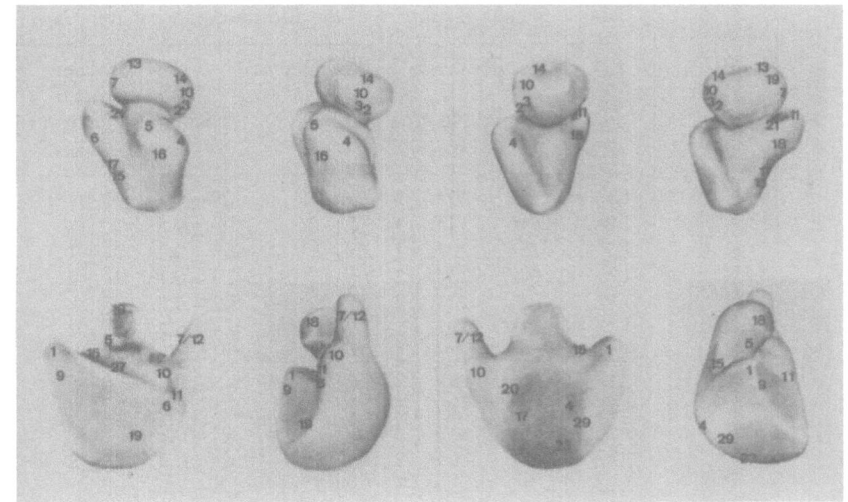

Fig. 8. Immuno electron microscopy: presentation of the small (upper row) and the large ribosomal subunit (lower row). The numbers mean the antigenic determinants of the corresponding ribosomal proteins. According to M. and G. Stöffler

ribosome, as well as on the ribosomal structure in various functional states. Such information should be collected under physiological conditions, for example, with an optimal buffer. These requirements are fulfilled by a new technique which has been developed for structural research on ribosomes during the last 15 years, namely the application of neutron scattering.

The basis for the use of neutron scattering in biology resides in the fact that more than 99.9% of all atoms present in the cell are represented by only six elements (H, C, N, O, P and S), and that the most frequent isotopes of these elements scatter neutrons in a very similar way, with the only exception being hydrogen. In constrast to hydrogen the heavy isotope deuterium (^{2}H) scatters like the other elements. The biological application of neutrons relies on this scattering difference between ^{1}H and the other elements including the heavy isotope of hydrogen ^{2}H. for example, if one integrates a deuterated component into a protonated multicomponent complex, then the deuterated component is "stained" for the neutron beam. This technique harbours two important advantages: 1) the biological activity of the complex is not affected by the integration of a deuterated component, and 2) the deuterated component is homogeneously stained due to the high hydrogen content of organic molecules.

Since the hydrogen density in H_2O is larger than that in biological molecules, the scattering densities of H_2O and D_2O represent extreme values. For this reason it is useful and usual that the scattering density of a biological sample is compared with the equivalent density of a D_2O/H_2O mixture (see Fig. 9). Thus, the

scatter of proteins is equivalent to a 40% D_2O mixture, and that of nucleic acids (RNA) to that of a 70% D_2O mixture. The different scattering densities of proteins and RNA can be traced back to their different hydrogen content. 50% of the atoms of protein are hydrogen, but only 37% of the atoms are hydrogen in nucleic acids such as RNA.

Ribosomes are heterogeneous for a neutron beam, as a result of the different scattering densities of proteins and RNA. This heterogeneity generates a significant background which impairs the complete evaluation of the scattering picture. We developed the following strategy to circumvent this disadvantage. Both RNA and protein moieties of the ribosome were differentially deuterated, so that the scattering density of each moiety corresponded to that of a 100% D_2O solution. A ribosome consisting of these deuterated RNA and protein components is homogeneous for the neutron beam. If such a ribosome is transferred into a water milieu containing 100% D_2O, then the ribosome

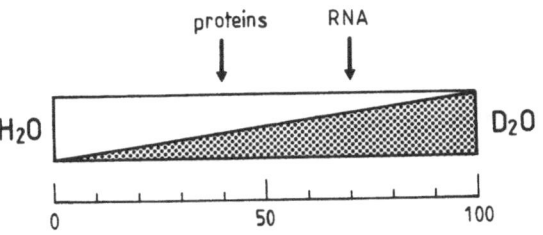

Fig. 9. Scattering density. The scattering densities of biological samples can be adjusted by mixing H_2O and D_2O

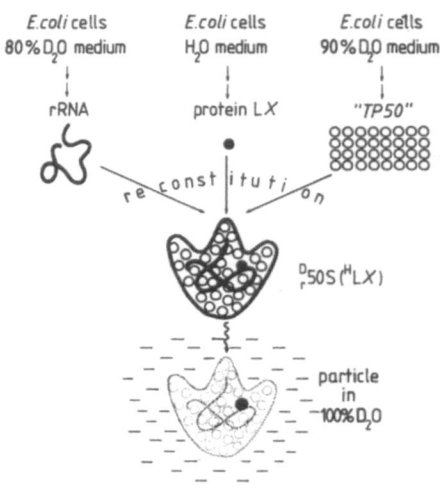

Fig. 10. Strategy for the production of *E. coli* ribosomes which are homogeneous for the neutron beam

Table 2. Shapes of proteins of the 50 S ribosomal subunit

	Molecular mass		Protein in situ		Isolated protein
	Sequence M_R	Neutron $M_R/1000$	Neutron R_g	Axial ratio	Neutron R_g
L1	24,599	20.8 ± 2	26 ± 2	4.9 ± 0.7	25 ± 1.5
L2	29,416	33.8 ± 3	22 ± 2	3.3 ± 0.6	
L3	22,258	20.8 ± 2	25 ± 2	4.9 ± 0.7	24 ± 1
L4	22,087	20.8 ± 2	19 ± 2	2.9 ± 0.7	29 ± 2
L9	15,531	12.8 ± 2	23 ± 2	5.2 ± 0.8	
L13	16,019	16.6 ± 2	21 ± 2	4.4 ± 0.7	26 ± 2
L16	16,296	12.8 ± 2	13 ± 2	1.3 ± 1.0	25 ± 2
L17	14,364	15.4 ± 2	14 ± 2	5.8 ± 0.8	
L18	12,770	15.7 ± 2	22 ± 2	5.4 ± 0.8	20 ± 2
L20	13,366	13.8 ± 2	18 ± 2	3.7 ± 0.8	
L22	12,227	11.8 ± 2	21 ± 2	5.1 ± 0.8	23 ± 2
L23	11,013	11.3 ± 2	15 ± 2	2.8 ± 0.9	
L24	11,185	10.3 ± 2	19 ± 2	4.5 ± 0.8	21 ± 2

becomes "invisible" for the neutron beam. The differential deuteration of proteins and RNA was achieved in the following way. *E. coli* cells were fermented in a medium containing a defined amount of D_2O, so that all the nucleic acids produced by the bacterium have a scattering density of near to 100 % (see Fig. 10). These cells were broken, the ribosomes isolated, and the RNA obtained from the large ribosomal subunit. In a second fermentation the D_2O content of the medium was adjusted, so that now all the proteins produced by the bacteria had a scattering density of near to 100 %. Again the large ribosomal subunits were isolated from these cells and the proteins prepared. The large subunit was now reconstituted from both fractions, the deuterated RNA and the deuterated proteins, together with one or two protein(s) in a protonated state. The ribosomal matrix now is homogeneous, and the neutron beam recognizes only the one or two *protonated* component(s) which were integrated into the ribosome. This technique is termed the strategy of the "glassy ribosome", and has enabled us to measure the form of single proteins within the ribosomal subunit.

Table 2 compiles the shape data obtained so far. The shape parameter extracted was the radius of gyration, *Rg*, which is used to calculate the main axis of a prolate ellipsoid of revolution which scatters in a equivalent manner. These main axes approximate to spatial dimensions of the corresponding protein in the ribosome. It can be seen from Table 2 that only one protein has a nearly spherical shape (L16, axial ratio about 1:1), whereas most of the proteins have a relatively globular

structure (axial ratio of up to 5:1), only a few proteins being significantly more elongated. An interesting result is that two proteins, L4 and L16, undergo a dramatic structural change during the assembly of the ribosome. They show a large radius of gyration in the isolated state but have a more compact, globular structure within the ribosome.

When two protonated proteins are integrated in the deuterated ribosomal matrix, then the distance between the mass centres of gravity of these two proteins can be determined. The distances so far measured are presented in Table 3, and those distances are used by

Table 3. Protein distances within the 50 S ribosomal subunit

Protein pair	Distance (Å)	Protein pair	Distance (Å)
L1–L2	75	L4 –L16	80
L1–L3	40	L4 –L21	160
L2–L3	120	L4 –L23	45
L2–L4	115	L4 –L24	60
L2–L17	105	L9 –L20	55
L2–L22	48	L9 –L24	135
L3–L4	Neighbours	L13–L20	90
L3–L9	140	L13–L23	95
L3–L13	55	L15–L20	135
L3–L15	100	L16–L24	100
L3–L18	155	L20–L22	80
L3–L21	160	L20–L23	75
L3–L22	160	L20–L24	95
L3–L23	110	L21–L24	95
L3–L24	100	L22–L23	80
L4–L9	140		

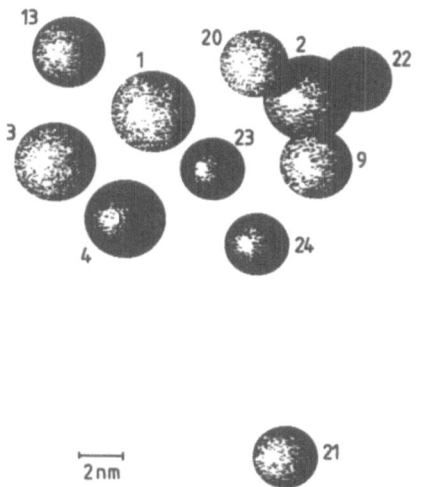

Fig. 11. A first preliminary spatial presentation of 11 proteins (mass centers of gravity) derived from the large subunit. The volumes of the spheres correspond to their molecular weight.

means of a geometric triangulation procedure to reconstruct the spatial distribution of the mass centres of gravity. The first preliminary spatial distribution of the mass centres of gravity of 11 proteins is shown in Figure 11. The handedness of the pattern is still undetermined. The volume of each sphere corresponds to the molecular weight of the corresponding protein.

The possibilities of the neutron scattering analysis are by no means exhausted with these structural aspects. The "glassy ribosome" can also be used to visualize the path of the messenger RNA within the ribosome, as well as to pursue the migration of a tRNA molecule through the ribosome during protein synthesis. The planned experiments could lead to a "cinematographic" description of protein biosynthesis.

3.3 The functions of the ribosome

The more than 20 testable partial functions of the ribosome can be assigned to three functional phases, namely the start of protein synthesis (initiation), the prolongation of the polypeptide by one amino acid (elongation), and finally the stop of protein synthesis (termination). During initiation the small ribosomal subunit searches for the start signal on the mRNA and is then complement with the large ribosomal subunit. The ribosome now enters the elongation phase. In one cycle of reactions one amino acid is selected in accordance with the codon information, and is added to the already synthesized polypeptide. If at the end of poly-

peptide synthesis a stop codon enters the ribosome, then the synthesized polypeptide is released and folding to the active protein structure is completed. The ribosome leaves the mRNA and dissociates into its subunits. The small subunit is then ready to initiate the synthesis of another protein.

During the last 20 years the ribosomal activities have been described in terms of a model in which there are two binding sites for tRNA. One binding site is the "P site" which is favoured by the tRNA with the already synthesized polypeptide chain. The second one is the "A site" for the newly incoming aminoacyl-tRNA. Figure 12, left half, illustrates the elongation cycle according to this two-site model. The elongation cycle was reduced to three essential reactions: 1. The binding of aminoacyl-tRNA according to the codon present at the A site; 2. The peptidyl transfer, where the polypeptidyl residue is transferred from the peptidyl-tRNA to the newly incoming aminoacyl-tRNA via a peptide bond, resulting in a peptidyl-tRNA now prolonged by one amino acid and located in the A site. 3. The translocation of the tRNA-mRNA complex across the ribosome by one triplet length, with the result that deacylated tRNA leaves the P site, the peptidyl-tRNA moves from A to P site and a new codon enters the A sites.

This two-site model relies on the enzymatic activity of the ribosome to form a peptide bond. This ability is tested with the so-called puromycin reaction. The P site is defined as the site from which the peptidyl residue can be transferred to an aminoacyl residue or to puromycin, an aminoacyl-tRNA analogue. The peptidyl-tRNA, however, is by definition present at the A site if this reaction does not occur.

Since the puromycin reaction only checks the acyl residue but not the tRNA moiety, we started some years ago to characterize the tRNA binding sites systematically by means of binding and titration experiments with tRNA. To our surprise a third binding site in addition to the A and P sites was found. This additional binding site could bind deacylated-tRNA in a codon specific manner. We termed this third binding site *E* site (*E* for exit) following a suggestion made 20 years ago. This suggestion was based on a result which hinted towards a third binding site in eukaryotic ribosomes, but at that time, this hint was not pursued further. We were able to demonstrate by titration experiments that deacylated-tRNA occupies the three binding sites in the sequence P, then E and finally A site.

The fact that the ribosome contains three tRNA binding sites, together with the hint from our functi-

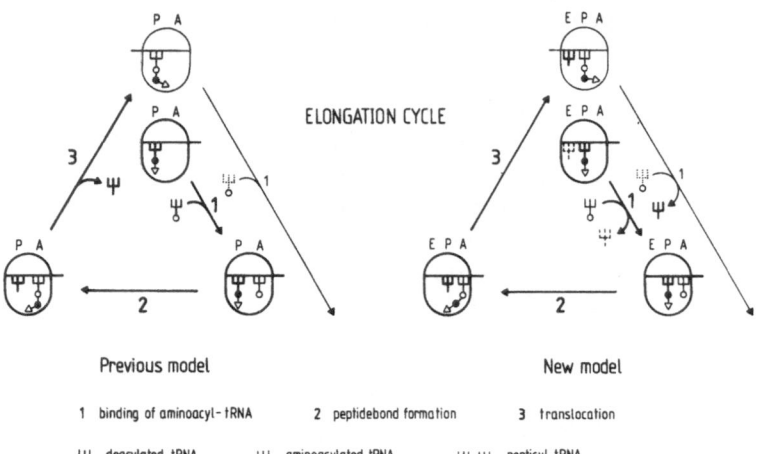

Fig. 12. The elongation cycle. Left half: previous two-site model. Right half: new three-site model

onal studies that in all functional states the ribosome carries at least two tRNAs, led to the proposal of a new three-site model (Fig. 12, right half). An essential difference between the two-site and the three-site models concerns the release of deacylated tRNA. In the two-site model the release is coupled to the translocation, whereas in the three-site model the translocation does not lead to a release of deacylated tRNA from the ribosome but rather to a movement of the deacylated tRNA from P to E site. The subsequent occupation of the A site should then be the trigger for the release of the deacylated tRNA from the E site.

We could indeed demonstrate that translocation is not coupled to tRNA release, but that instead the A site occupation leads to release of deacylated tRNA. Occupation of the E-site in turn lowers the affinity at the A site. Obviously, A and E sites are allosterically linked in a bidirectional manner. Occupation of one site reduces the affinity for the tRNA at the other site (negative cooperativity).

Thus, the ribosomal functions during protein biosynthesis are characterized by the following features: 1. The ribosome contains three tRNA binding sites,

the A, P and E sites. 2. The allosteric link between A and E sites (negative cooperativity) has the consequence that the ribosome always contains two high affinity sites: before translocation these sites are A and P, and after translocation P and E. 3. Both tRNAs on the ribosome are continuously connected to the mRNA via codon-anticodon interaction (see Fig. 13).

The three-site model has further important consequences concerning both precision of aminoacyl-tRNA selection and the translocation reaction. 1. The fact that an occupied E site lowers the A site affinity means that codon-anticodon interaction, which is only a part of tRNA-ribosome interaction, represents a significant energy increment for the binding to the A site. It follows that only those aminoacyl-tRNAs will take part in the binding reaction, which harbour the precisely corresponding anticodon (cognate tRNA) or at least a very similar anticodon (miscognate tRNA). The remaining 90 % of the aminoacyl-tRNAs which contain an incorrect anticodon (non-cognate tRNAs) are not bound. Therefore, the non-cognate tRNAs do not interfere with protein synthesis, and the ribosome "only" needs to distinguish the tRNAs having the pre-

three-site model: allosteric interaction between A and E

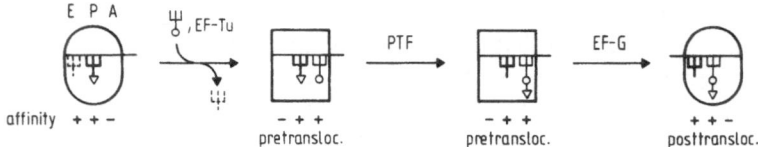

Fig. 13. The allosteric three-site model for the description of the ribosomal elongation. EF-Tu and EF-G are elongation factors which are involved in the aminoacyl-tRNA binding to the A site and the translocation, respectively

cisely fitting anticodon from those having a very similar one, by means of a mechanism (proof-reading mechanism) which is not yet unterstood. 2. With regard to the translocation reaction the two adjacent codon-anticodon interactions establish a rather tight connection between both tRNAs on one hand and the mRNA on the other hand. This tight connection might be important for keeping the reading frame during the translocation reaction. 3. After translocation, both codon-anticodon interactions at P and E sites may be important for a precise exposition of the new codon in the A site. A precise exposition of the codon is essential for a high accuracy of aminoacyl-tRNA selection at the A site.

The role of the third site for the initiation and termination phase has still to be worked out.

Acknowledgements

I thank Dr. H. G. Wittmann for continuous interest. I am indebted to Dr. R. Brimacombe for many discussions and advice, to E. Philippi for the drawings, and to Drs. P. and V. Nowotny for their help in preparing the manuscript. I am grateful to my coworkers A. Bartetzko, U. Geigenmüller, T.-P. Hausner, M. Herold, C. Jantz, P. Nowotny, V. Nowotny, H.-J. Rheinberger, S. Schilling, C. Stiege, P. Tatzel, H. Voß and P. Wurmbach; it is a pleasure to analyze with them some aspects of the challenging ribosome.

References

1. Alberts B, Bray D, Lewis J, Raff M, Roberts K, Watson JD (1983) Molecular Biology of the Eucaryotic Cell. Garland Publishing, New York, Chapters 1–3
2. Nierhaus KH (1982) Structure, Assembly and Function of Ribosomes. Curr Topics Microbiol Immun 97:82–155
3. Nomura M, Held W (1974) (eds) Nomura M et al, Ribosomes, cold Spring Harbor Laboratory, New York, pp 393–416
4. Nierhaus KH (1980) (eds) Chambliss G et al, Ribosomes, Structure, Function and Genetics, University Park Press, Baltimore pp 267–294
5. Lake JA (1981) Sci Am 245:84–97
6. Stöffler G, Stöffler-Meilicke M (1986) (eds) Hardesty B, Kramer G, Ribosomes, Springer Verlag, New York, in press
7. Moore PB (1980) (eds) Chambliss G et al, University Park Press, Baltimore, pp 111–133
8. Nowotny V, May RP, Nierhaus KH (1986) (eds) Hardesty B, Gramer G, Ribosomes, Springer Verlag, New York, in press
9. Nierhaus KH, Rheinberger HJ (1984) TIBS 9:428–432
10. Specific features which have not been mentioned in this article (primary sequence of proteins, secondary structure of rRNA, genetics and regulation of ribosomal components, eucaryotic ribosomes etc) can be found in Chambliss G et al (eds) (1980) Ribosomes, Structure, Function and Genetics, University Park Press, Baltimore

Received December 12, 1985;
accepted December 19, 1985

Author's address:

Knud H. Nierhaus
Max-Planck-Institut für Molekulare Genetik
Abteilung Wittmann
Ihnestraße 63–73
D-1000 Berlin 33 (Dahlem), F.R.G.

Lipopeptides, an attractive class of microbial surfactants*)

J. Vater

Institut für Biochemie und Molekulare Biologie, Technische Universität Berlin, F.R.G.

Abstract: Microorganisms produce a broad spectrum of biosurfactants which receive increasing attention for industrial, biotechnological and therapeutical applications. Lipopeptides are a class of microbial surfactants which are of particular interest because of their wide range of activities. A short survey is given on their occurrence, structure and function. Various strains of Bacillus subtilis produce a family of lipopeptides which are powerful antibiotics with antifungal activities. A representative of these agents is surfactin which is one of the most efficient biosurfactants so far known. It has been purified by several chromatographic procedures. In vivo incorporation experiments with ^{14}C-labelled precursor amino acids have been performed. It is shown that the biosynthesis of this agent appears in the logarithmic phase of bacterial growth and continues over a wide range of the cell cycle.

Key words: Microbial surfactants, lipopeptides, biosynthesis.

Introduction

Microorganismus produce a broad spectrum of extracellular biosurfactants [1] which are released into the culture medium. Many of such agents are lipid containing compounds. Their surface active properties result from a combination of polar and apolar structural elements in the surfactant molecule. The apolar, hydrophobic part is frequently a hydrocarbon chain. The polar components appear in manifold variations. They comprise sugars, amino acids, phosphates or alcohol and ester groups in the case of neutral lipids.

Such compounds are receiving increasing attention for industrial, biotechnological and therapeutical uses. Their utilization seems attractive, because these biosurfactants show a broad spectrum of molecular structure and surface activities. In particular, the suitable agent for a selected application may be adapted by the choice of the producer organism, the composition of the nutrient broth and the culture conditions. Furthermore such biosurfactants often are less toxic and better biodegradable than synthetic compounds. In this way the potential of pollution may be minimized.

Many of these compounds are effective emulsifiers. Their utilization for enhanced oil recovery or the clean-up of oil spills is under current research [2] and has been discussed at various international conferences [3, 4]. Such biosurfactants facilitate the growth of their producer organisms on hydrocarbon substrates. In this way waste products can be converted to valuable materials, as single cell protein, sugars, polysaccharides etc. These aspects qualify naturally occuring surfactants for an interesting field of research.

Lipopeptides, an attractive class of microbial surfactants — a short survey

Lipopeptides represent a class of microbial surfactants which attain increasing scientific, therapeutical and biotechnological interest. A collection of such agents is shown in Table 1. Their occurrence and activities are indicated. Their structures are demonstrated in Table 2.

Such compounds are found wide-spread over the whole spectrum of microorganisms. They frequently

---

*) Lecture presented during the 32nd Annual Meeting of the Kolloid-Gesellschaft, Berlin October 2–4, 1985.

Table 1. Lipopeptides produced by various microorganisms

Name	Producer organism	Properties and activities
1. Amphomycin [5]	Streptomyces canus	Antibiotic, inhibitor of cell wall synthesis
2. Chlamydocin [6,7]	Diheterospora chlamydosporia	Cytostatic and antitumor agent
3. Cyclosporin A [8,9]	Tolypocladium inflatum (Trichoderma polysporum)	Antifungal agent, immunomodulator
4. Enduracidin A [10–13]	Streptomyces fungicidicus	Antibiotic
5. Globomycin [14]	Streptomyces globocacience	Antibiotic, inhibitor of cell wall synthesis
6. HC-Toxin [15–17]	Helminthosporium carbonum	Phytotoxin
7. [18]	Microcystis aeruginosa (cyanobacterium)	Toxin, induces gastroenteritis, dermatitis and liver damage in human subjects
8. Polymyxin E₁ (Colistin A) [19]	Bacillus polymyxa	Antibiotic

Table 2. Structures of the lipopeptides compiled in Table 1

1. FA → Asp → MeAsp → Asp → Gly → Asp → Gly → Dab^e → Val → Pro → Dab^t → Pip
    FA = (+) -3-anteisotridecenoic or 3-isotridecenoic acid

2. cylco (L-AOE → Aib → Phe → Pro)

3.   ↗ FA —— Abu —— Sar —— MeLeu —— Val
    MeVal
         ↖ MeLeu ← MeLeu ← D-Ala ← Ala ← MeLeu
    FA = 2-methylamino-3-hydroxy-4-methyl-6-octenoic acid

4. FA → Asp → Thr → D-HyPhg → D-Lys → D-aThr → HyPhg → D-HyPhg → aThr → Cit
         HyPhg ← D-Ala ← End ← Gly ← D-CHyPhg ← D-Ser ← HyPhg ← D-End ←
    FA = 10-methyl-undeca-2-(cis), 4(trans)-dienoic acid

5. MeLeu → aIle → Ser → aThr → Gly
       ↑
       CO
       |
    CH₃–CH ———— CH–O←
           |
           R
    R = –(CH₂)₅–CH₃

6. cyclo (L-AOE → D-Pro → Ala → D-Ala)

7.   ↗ D-Glu → Me Δ Ala → D-Ala
    Adda
         ↖ Ala ← β-MeAsp ← Leu

8. 6-CH₃-octanoyl → Dab → Thr → Dab → Dab → D-Leu → Leu
            ↑
            Thr ← Dab ← Dab ←

Dab = 2,4-diaminobutyric acid; Dab^e = D-erythro-2,3 diaminobutyric acid; Dab^t = L-threo-2,3-diaminobutyric acid; Pip = D-α-pipecolic acid; Aib = 2-aminoisobutyric acid; Abu = 2-aminobutyric acid; Sar = sarcosin; HyPhg ; 2-amino-4-hydroxyphenylacetic acid or hydroxyphenylglycine; CHyPhg = 3-chloro- or 3,5-dichloro-HyPhg; Cit = citrulline; End = enduracidine or α(s)-amino-β-4 (R)-(2-imino imidazolidinyl)-propionic acid; aThr = allo-threonine; aIle = allo-isoleucine; MeΔAla = N-Methyl-dehydroalanine; β-MeAsp = erythro-β-methyl-D-aspartic acid; AOE = 2-amino-9, 10-epoxy-8-oxodecanoic acid; Adda = 3-amino-9-methoxy-10-phenyl-2,6,8-trimethyl-deca-4,6-dienoic acid.

occur in bacteria and fungi. Recently a lipopeptide has also been isolated from the cyanobacterium Microcystis aeruginosa [18]. From intensive screening of other blue-green algae the isolation of similar compounds can be expected. The characteristic structural element of such lipopeptides is a sepcific fatty acid which is combined with an amino acid moiety. Here rare and modified amino acids are found as constituents which are not used for ribosomal protein synthesis. Such bioactive peptides usually appear as mixtures of closely related compounds which show slight variations in their amino acid composition and/or in their lipid portion.

The spectrum of the manifold activities of lipopeptides covers antibiotics, antifungal, antiviral and antitumor agents, immunomodulators or specific toxins and enzyme inhibitors. Though the mechanism of action of the majority of such compounds has not been clarified in detail so far, it is obvious that their surface — and membrane-active properties play an important role in the expression of their activities. Therefore, the analysis of their physicochemical parameters and molecular structures as well as the identification of their targets in the living cell are prerequisites for the elucidation of their mode of action. Our research concentrates on a class of lipopeptides which are formed by certain

strains of Bacillus subtilis, in particular, surfactin [20–26] and bacillomycin L [27, 28]. These compounds represent a family of cyclic peptides which consist of 8–17 amino acids. Some of these agents are listed in Table 3.

With the exception of mycobacillin they are distinguished by a β-amino- or β-hydroxy-C_{14}-C_{16} fatty acid as characteristic structural element. These components appear as mixtures of closely related isomers. For example, the lipid moiety of bacillomycin is composed of the following β-amino fatty acids [28]: n-C_{14} (38.9 %); iso-C_{15} (25.2 %); anteiso-C_{15} (15.4 %); iso-C_{16} (10.1 %) and n-C_{16} (6.1 %). The main fatty acid constituent of surfactin is 3-hydroxy-13-methyltetradecanoic acid [25, 34].

The B. subtilis lipopeptides are powerful antibiotics with antifungal activities. They show an exceptionally high surface activity. Surfactin is one of the most efficient biosurfactants so far known. In an aqueous solution which contains 0.005 % of this agent the surface tension of water is decreased from 72 dyn/cm to approx. 27 dyn/cm [20].

Surfactin inhibits fibrin clot formation and lyses erythrocytes and several bacterial spheroplasts and protoplasts [20, 35]. It functions as a potent inhibitor of cyclic adenosine 3′,5′-monophosphate phosphodiesterase [26].

Most of such bioactive peptides are synthesized by their producer microorganisms nonribosomally [36–41]. Two enzymatic principles have so far been detected for the synthesis of such compounds. Characteristic examples of the first reaction principle are the peptide antibiotics produced by B. brevis (gramicidin S and tyrocidines). They are formed on multifunctional polypeptide chains which activate their substrate amino acids in a two-step mechanism involving aminoacyl adenylates and thioesters. Multifunctional peptide synthetases of this type are equipped with a 4′-phosphopantheine swinging arm as an internal transport system. The elongation of the growing peptide chain proceeds in a series of transpeptidation and transthiolation steps by interactions of the central SH-group of the carrier with the peripheral thiols at the reaction centers of the multienzyme (Thiotemplate-mechanism). The intermediate peptides are covalently attached to the producer enzymes. They are vectorially transported within the enzyme structure and cannot be exchanged with the reaction medium. The sequence of such a peptide is coded by the structural organization of the activation domains on the multifunctional polypeptide chain.

Table 3. Lipopeptides produced by various strains of Bacillus subtilis

Surfactin [20–26]

$$L\text{-Leu} \rightarrow \beta\text{-OH-}C_{13-15} \rightarrow L\text{-Glu} \rightarrow L\text{-Leu}$$
$$\uparrow \qquad\qquad\qquad\qquad\qquad\qquad \downarrow$$
$$D\text{-Leu} \leftarrow L\text{-Asp} \leftarrow L\text{-Val} \leftarrow D\text{-Leu}$$

Bacillomycin L [27, 28]

$$L\text{-Thr} \rightarrow \beta\text{-N-}C_{14-16} \rightarrow L\text{-Asp} \rightarrow D\text{-Tyr}$$
$$\uparrow \qquad\qquad\qquad\qquad\qquad\qquad \downarrow$$
$$D\text{-Ser} \leftarrow L\text{-Gln} \leftarrow L\text{-Ser} \leftarrow D\text{-Asn}$$

Iturin A [29]

$$L\text{-Ser} \rightarrow \beta\text{-N-}C_{14-16} \rightarrow L\text{-Asn} \rightarrow D\text{-Tyr}$$
$$\uparrow \qquad\qquad\qquad\qquad\qquad\qquad \downarrow$$
$$D\text{-Asn} \leftarrow L\text{-Pro} \leftarrow L\text{-Gln} \leftarrow D\text{-Asn}$$

Mycosubtilin [30]

$$\nearrow L\text{-Asn} \rightarrow \beta\text{-N-}C_{16}, C_{17} \rightarrow L\text{-Asn} \rightarrow L\text{-Gln}$$
$$D\text{-Ser} \qquad\qquad\qquad\qquad\qquad\qquad\qquad \downarrow$$
$$\searrow D\text{-Asn} \leftarrow D\text{-Asn} \leftarrow D\text{-Tyr} \leftarrow L\text{-Pro}$$

Mycobacillin [31–33]

$$L\text{-Tyr} \rightarrow L\text{-Asp} \rightarrow L\text{-Tyr} \rightarrow L\text{-Ser} \rightarrow D\text{-Asp} \rightarrow L\text{-Leu} \rightarrow D\text{-Glu}$$
$$\uparrow D\text{-Glu} \leftarrow D\text{-Asp} \leftarrow L\text{-Pro} \leftarrow D\text{-Asp} \leftarrow L\text{-Ala} \leftarrow D\text{-Asp} \leftarrow$$

Representative for the second reaction principle is the biosynthesis of glutathione, a linear tripeptide γ-Glu-Cys-Gly [42]. Here carboxyl transfer proceeds from phosphate activated intermediates in an one-step mechanism. If the elongation process is interrupted by omission of substrate amino acids, intermediate peptides can be released into the reaction medium. Glutathione is formed by the cooperation of two freely interacting or weakly associated enzymes. Similar reaction elements have been reported by Bose and his collegues [31–33] for the biosynthesis of mycobacillin, a cyclic tridecapeptide produced by B. subtilis B 3. These authors studied this process using a cell free system of this organism [33]. At present, however, no information is available on the nature of the multienzyme systems involved in the formation of this peptide antibiotic and related compounds in B. subtilis.

Our research, therefore, concentrates a) on the isolation, purification and characterization of the synthetases for these lipopeptides and b) on the analysis of the structural and functional organization of the producer multienzymes as well as of the code that determines the sequence of these agents.

Experimental section

Materials

[14]C-L-leucine and ^{14}C-L-aspartic acid were purchased from Amersham Buchler (Braunschweig, F.R.G.). Acetonitrile (Lichrosolv grade) and silica gel thin layer plates were products of Merck (Darmstadt, F.R.G.). Column materials were LH-20 and DEAE Sephacel (Pharmacia, Freiburg, F.R.G.).

Methods

Growth of organism:

Bacillus subtilis ATCC 21332 was cultivated in Erlenmeyer flasks at 28 °C in the medium introduced by Landy et al. [43] using a New Brunswick shaker G-25. The rotation frequency was 120 rpm. Cell growth was monitored by turbidimetry at 650 nm at intervals of 1 h. At the same time the pH of the cell culture was controlled.

Isolation and detection of surfactin:

Bacillus subtilis cells were removed from the cell culture by centrifugation. Surfactin was precipitated from the culture medium by addition of 1 N HCl, collected by centrifugation and prepurified by procedures reported by Bernheimer and Avigad [35] and Cooper et al. [44]. The rough material was extracted with dichloromethane. The solvent was removed in vacuo. The remaining solid was dissolved in distilled water. Sufficient NaOH was added to adjust a pH between 7–8. Surfactin was precipitated by addition of solid NaCl adjusting a concentration of 10 % (w/v). The precipitate was

collected by centrifugation. At this stage of purification surfactin was detected either by thin layer chromatography on silica gel sheets, as reported by Cooper et al. [44] or by inhibition of the growth of Penicillium chrysogenum tested by the cylinder agar plate diffusion method [45]. Radioactively labelled surfactin was monitored either by liquid scintillation counting or by thin layer chromatography and radioscanning. Using the latter technique aliquots of fractions were spotted on silica gel sheets (Merck, D-60, 20 × 20 cm). Chloroform/methanol/water (65:25:4) was used as mobile phase. ^{14}C-surfactin was detected using a Berthold Linear Analyzer LB 2832.

Results and discussion

Prerequisites for in vitro studies of the biosynthesis of microbial peptides using the isolated multienzyme systems are:

a) the development of techniques for the purification and analytical determination of the peptide products and

b) in vivo incorporation experiments with radioactively labelled precursors. Such studies yield limited information on the mechanism of peptide formation and define the period in the cell cycle of the producer organism during which the peptide is synthesized.

In this publication we report on such research concerning the biosynthesis of surfactin.

1. In vivo incorporation of radioactively labelled precursor amino acids into surfactin

Bacillus subtilis ATCC 21332 was grown in the Landy medium, as indicated in the Experimental section. A growth curve of this organism is shown in Figure 1. After inoculation there is a short burst of bacterial growth during the first 5 hours, presumably due to the consumption of endogeneous substrates. Then after a lag phase of approx. 5 hours the bacillus enters the logarithmic phase of growth. Now within 8 hours a tenfold increase of the optical density at 650 nm of the cell suspension is observed. Finally a period of a slow linear progress of the OD appears. Sporulation of B. subtilis occurs in the range of 45–60 hours after the start of fermentation. During this process the pH increases from 5.8 to 7.0.

Biosynthesis of surfactin was monitored by in vivo incorporation of two of its precursor amino acids. 1 μCi of either ^{14}C-L-leucine or ^{14}C-L-aspartic acid was added to a series of 500 ml Erlenmeyer flasks which contained 100 ml of the growing culture of the B. subtilis cells at 3 characteristic points of the bacterial growth (10 h, 20 h and 30 h after the start of fermentation). The

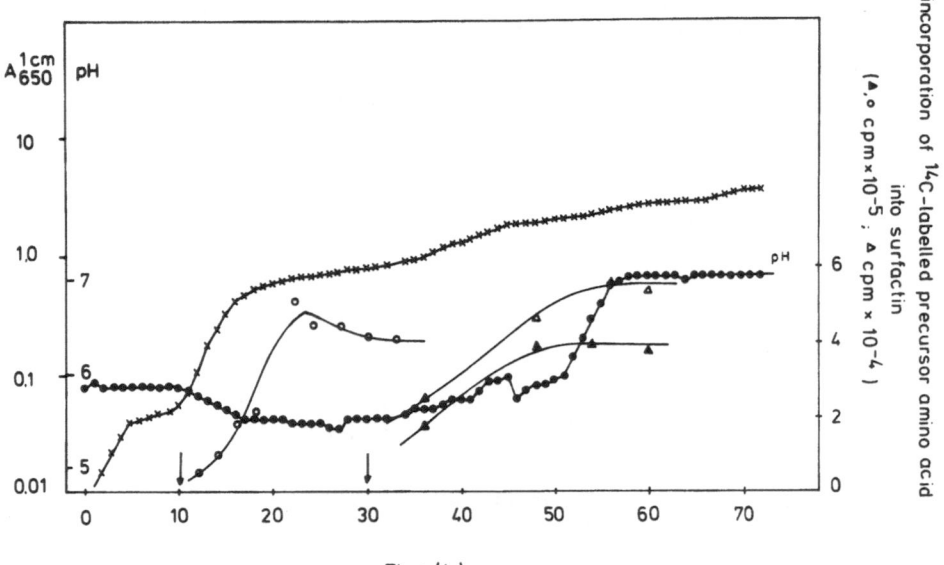

Fig. 1. Growth of Bacillus subtilis ATCC 21332 at 28 °C in the Landy medium. Bacterial growth as monitored by the turbidity of the culture at 650 nm (×) and the pH of the medium (●) were measured at intervals of 1 h. Each 1 µCi of a ^{14}C-labelled precursor amino acid was added to the growing culture at 10 and 30 h after the start of fermentation. Incorporation of ^{14}C-L-Leu (○, ▲) and ^{14}C-L-Asp (△) into surfactin was detected by scintillation counting at various intervals after addition of the tracer

radiolabelled product was harvested at various time intervals after the addition of the tracer. Surfactin was isolated from the culture medium by precipitation with 1 N HCl. The crude material was extracted with dichloromethane and further purified by salt precipitation with NaCl. After this procedure the product was dissolved in ethanol. Aliquots were analyzed by scintillation counting. Precursor incorporation was determined as a function of the time intervals between its addition to the bacterial culture and harvesting of the lipopeptide.

In the first experiment the tracer was added at the beginning of the bacterial growth (10 h after the start of fermentation). From Figure 1 it is apparent that surfactin production starts early in the logarithmic phase and proceeds until the substrate is quantitatively converted into the lipopeptide within a period of approx. 10 h. In the second experiment the precursor was added at the end of the logarithmic phase (20 h after the start of fermentation; data not shown). In this case ^{14}C-L-leucine was incorporated into surfactin at a high rate. Substrates were exhausted already after 5 hours. If feeding occurs in the period of the slow linear increase of optical density at 30 hours after the start of fermentation (third experiment) surfactin production is still active, but proceeds at an appreciably lower rate than in the previous experiments.

Obviously biosynthesis of surfactin appears in the logarithmic phase of growth and continues over a wide range of the cell cycle. Optimal rates of lipopeptide

formation were observed at the end of the logarithmic growth. Therefore, B. subtilis should be harvested in this period of the cell cycle in order to prepare an in vitro system which is needed for enzymatic studies of the biosynthesis of surfactin.

2. Purification of surfactin

Highly purified surfactin preparations were obtained by the application of a series of chromatographic procedures. After the NaCl precipitation the crude material was gel filtrated on a Sephadex LH-20 column (2 × 25 cm).

Hexane/chloroform/water (25 : 45 : 10) (v/v) was used as the eluent. Surfactin was monitored by the absorption at 220 nm, thin layer chromatography [44] and inhibition of the growth of Penicillium chrysogenum using the cylinder agar plate diffusion method [45]. In this purification step coloured impurities of surfactin were removed. The lipopeptide containing fractions were collected and the solvent removed in vacuo. The residue was dissolved in 10 mM Tris-HCl buffer, pH = 8,0 with addition of 10 % (v/v) ethanol and loaded on a DEAE Sephacel column (2 × 26 cm). Surfactin was eluted with a linear gradient from 0–1 M NaCl. The product appeared in a conductivity range of 20–30 mΩ^{-1} as a single peak (Fig. 2). Surfactin was precipitated from the active fractions with 0,1 N HCl and collected by centrifugation. The white pellet was washed with 0,1 N HCl, dissolved in 10 mM ammon-

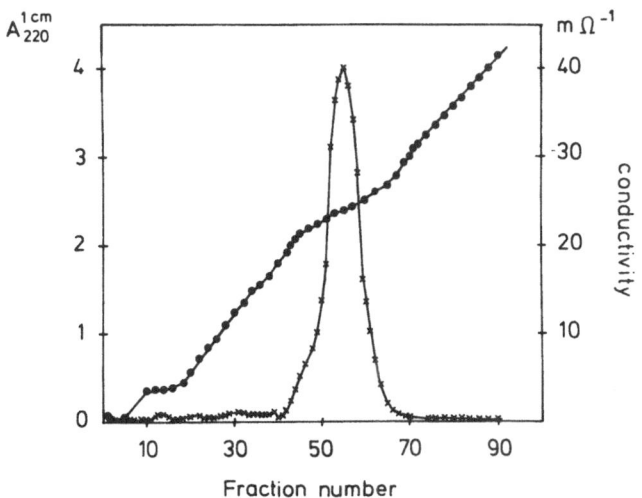

Fig. 2. Anion exchange chromatography of surfactin on DEAE-Sephacel. The lipopeptide was eluted with a linear gradient from 0 to 1 M NaCl and monitored by the absorption at 220 nm ($\times$). The conductivity of each fraction was measured ($\bullet$)

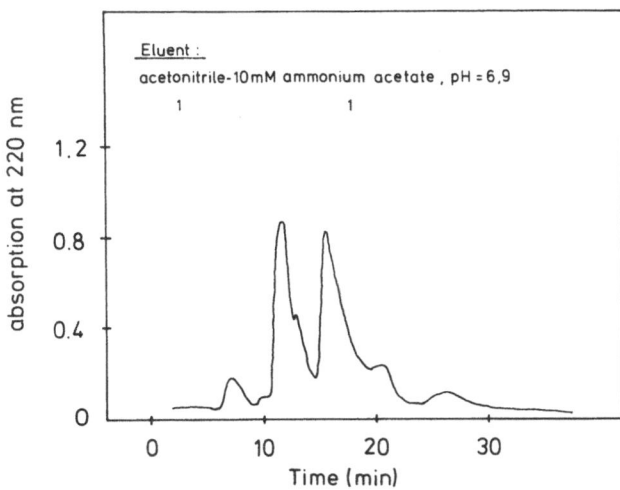

Fig. 3. Isocratic reversed phase HPLC of surfactin on RP 18 ODS Hypersil with acetonitrile — 10 mM ammonium acetate, pH = 6,9 as the eluent. Flow rate: 0,5 ml/min. The absorption was monitored at 220 nm. A LKB HPLC-system was used

ium acetate, pH = 6,9 and loaded onto a RP 18 ODS Hypersil column. The product was separated into at least 5 fractions by isocratic reversed phase HPLC, as demonstrated in Figure 3. Here a similar procedure was applied, as has been reported by Peypoux et al. [28] for the separation of the components of bacillomycin L. Acetonitrile — 10 mM ammonium acetate (1 : 1) was used as the eluent.

Chromatographic purification of surfactin was also performed with radioactively labelled material obtained by in vivo incorporation of ^{14}C-leucine. After the DEAE Sephacel step the lipopeptide appeared as a single peak on thin layer chromatograms as detected by radioscanning (Fig. 4). The R_F of the product was 0,65.

In the final HPLC pattern radioactivity was found in all peaks shown in Figure 3. Preliminary amino acid analysis data indicate that the lipopeptide material obtained from all these fractions shows an amino acid composition characteristic for surfactin (see Table 3).

From these results we assume that surfactin produced by B. subtilis ATCC 21332 also represents a mixture of a few closely related compounds which vary in their lipid portion. Obviously these components can be separated by reversed phase HPLC. Their identification by FAB-mass spectrometry and gas chromatography is in preparation.

Acknowledgements

This research was supported by the Deutsche Forschungsgemeinschaft (Sonderforschungsbereich 9 „Struktur, Funktion und Biosynthese von Peptiden und Proteinen"). The expert technical assistance of Mrs. B. Kluge is appreciated. I thank Professer J. Salnikow for amino acid analysis of surfactin.

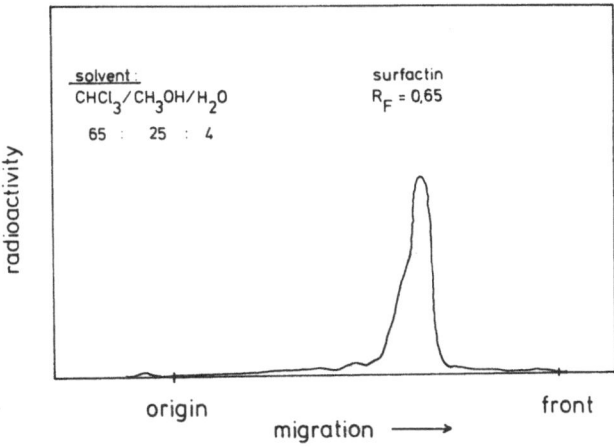

Fig. 4. Thin layer chromatography and radioscanning of ^{14}C-labelled surfactin. The material obtained after the DEAE-Sephacel chromatography was precipitated with 1 N HCl. The pellet was dissolved in ethanol. Aliquots were spotted on silica gel sheets (Merck DC-60, 20 × 20 cm). CHCl$_3$/CH$_3$OH/H$_2$O (65 : 25 : 4) was used as mobile phase. Radioscanning was performed using a Berthold Linear Analyzer LB 2832

References

1. Cooper DG, Zajic JE (1980) Adv Appl Microbiol 26:229
2. Finnerty WR, Singer ME (1983) Biotechnology 1:47
3. Department of Energy, Internat Conf on microbial processes useful in enhanced oil recovery (1979) CONF-790871, US Department of Energy, Washington, DC
4. National Science Foundation (1976) The role of microorganisms in the recovery of oil, Proc Engineering Foundation Conf NSF/RA 770201, National Science Foundation, Washington DC
5. Bodanszky M, Sigler GF, Bodanszky A (1973) J Am Chem Soc 95:2352
6. Closse A, Huguenin R (1974) Helv Chim Acta 57:533
7. Kawai M, Jasensky RD, Rich DH (1983) J Am Chem Soc 105:4456
8. Dreyfuss M, Härri E, Hofmann H, Kobel H, Pache W, Tscherter H (1976) Eur J Appl Microbiol 3:125
9. Rüegger A, Kuhn M, Lichti H, Lossli H-R, Huguenin R, Quiquerez C, von Wartburg A (1976) Helv Chim Acta 59:1075
10. Higashide E, Hatano K, Shibata M, Nakazawa K (1968) J Antibiot 21:126
11. Asai M, Muroi M, Sugita N, Kawashima H, Mizuno K, Miyake A (1968) J Antibiot 21:138
12. Hori M, Iwazaki H, Horii S, Yoshida I, Hongo T (1973) Chem Pharm Bull 21:1175
13. Iwazaki H, Horii S, Asai M, Mizuno K, Ueyanagi J, Miyake A (1973) Chem Pharm Bull 21:1184
14. Inukai M, Enokita R, Torikata A, Nakahara M, Iwado S, Arai M (1978) J Antibiot 31:410
15. Walton JD, Earle ED, Gibson BW (1982) Biochem Biophys Res Commun 107:785
16. Gross ML, McCrery D, Crow F, Tomer KB, Pope MR, Ciufetti LM, Knoche HW, Doly JM, Dunkle LD (1982) Tetrahedron Lett 23:5381
17. Pope MR, Ciufetti LM, Knoche HW, McCrery D, Daly JM, Dunkle LD (1983) Biochemistry 22:3502
18. Botes DP (1985) 13th Internat Congr Biochem, Amsterdam, Abstract Mo-218
19. Storm RD, Rosenthal KS, Swanson PE (1977) Ann Rev Biochem 46:723
20. Arima K, Kakinuma A, Tamura G (1968) Biochem Biophys Res Commun 31:488
21. Kakinuma A, Hori M, Isono M, Tamura G, Arima K (1969) Agric Biol Chem 33:971
22. Kakinuma A, Sugino H, Isono M, Tamura G, Arima K (1969) Agric Biol Chem 33:973
23. Kakinuma A, Hori M, Sugino H, Yoshida I, Isono M, Tamura G, Arima K (1969) Agric Biol Chem 33:1523
24. Hosono K, Suzuki H (1983) J Antibiot 36:667
25. Hosono K, Suzuki H (1983) J Antibiot 36:674
26. Hosono K, Suzuki H (1983) J Antibiot 36:679
27. Besson F, Peypoux F, Michel G, Delcambe L (1977) Eur J Biochem 77:61
28. Peypoux F, Pommier M-T, Das BC, Besson F, Delcambe L, Michel G (1984) J Antibiot 37:1600
29. Peypoux F, Guinand M, Michel G, Delcambe L, Das BC, Lederer E (1978) Biochemistry 17:3992
30. Peypoux F, Michel G, Delcambe L (1976) Eur J Biochem 63:391
31. Majumdar SK, Bose SK (1960) Biochem J 74:596
32. Sengupta S, Bose SK (1972) Biochem J 128:47
33. Sengupta S, Bose SK (1971) Biochim Biophys Acta 237:120
34. Kakinuma A, Ouchida A, Shima T, Sugino H, Isono M, Tamura G, Arima K (1969) Agric Biol Chem 33:1669
35. Bernheimer AW, Avigad LS (1970) J Gen Microbiol 61:361
36. Lipmann F (1973) Acc Chem Res 6:361
37. Laland SG, Zimmer T-L (1973) Essays Biochem 9:31
38. Katz E, Demain AL (1977) Bacteriol Rev 41:449
39. Kleinkauf H, Koischwitz H (1978) Progr Mol Subcell Biol 6:59
40. Kleinkauf H, von Döhren H (1981) Current Topics Microbiol Immunol 91:129
41. Kurahashi K (1981) (ed) Corcoran JW, Antibiotics, Vol IV, Springer, Berlin Heidelberg New York, p 325
42. Meister A, Tate S (1976) Ann Rev Biochem 45:559
43. Landy M, Warren GH, Rosenman SB, Colio LG (1948) Proc Soc Exp Biol Medicine 67:539
44. Cooper DG, Mac Donald CR, Duff SJB, Kosaric N (1981) Appl Environm Microbiol 42:408
45. Besson F, Peypoux F, Michel G, Delcambe L (1978) J Antibiot 31:284

Received December 3, 1985;
accepted May 15, 1986

Author's address:

Priv.-Doz. Dr. Joachim Vater
Institut für Biochemie und Molekulare Biologie
Technische Universität Berlin
Franklinstr. 29
D-1000 Berlin 10, F.R.G.

Progress in Colloid & Polymer Science

Progr Colloid & Polymer Sci 72:19–27 (1986)

Colloid systems for depot dosage forms*)

I. Zimmermann

Schering AG, Berlin, F.R.G.

Abstract: The progress of pharmacokinetics led to a deeper understanding of drug action. We realized in particular the impact of dosage forms on the bioavailability of drugs.

Consequently it became more and more important to understand the mechanisms of drug release from dosage forms in order to design delivery systems with controlled release characteristics.

In this context colloid or polymer based systems are important.

Drug release from W/O-emulsion systems is diffusion-controlled. The inner phase acts as a depot as can be seen from release data showing two diffusion processes with different kinetics. Polydimethylsiloxane systems are in a rubbery state at room temperature. Under these conditions the free volume concept of diffusion can be applied. It is shown that diffusion coefficients can be adjusted by varying the relative network density. By an appropriate design of the release system constant drug release over long periods of time can be achieved.

Key words: Biopharmaceutics, dosage forms, drug release, W/O-emulsion, polydimethylsiloxane.

The last 80 years, particularly the time since the Second World War, have seen the introduction of a number of excellent new drugs. If we look at the different types of drugs more closely, especially the history of pharmacy, it is astonishing to learn that the Egyptians, for example, already knew most of the dosage forms familiar to us today. Eber's papyrus, going back 3,500 years, carries descriptions of syrups, pills, ointments, electuaries and bandages for the treatment of disease. Mothes introduced capsule products in France in 1833. Aerosols were described for the first time by Donnan around 1900. We do not know exactly when they were first used as an inhalant dosage form. However, they were used on a much wider scale once the relevant propellant gases were available in sufficiently high quality.

So although most dosage forms have been known for a long time, there has been a tremendous breakthrough in the development of dosage forms in the last few years and the end has by no means yet been reached. This development essentially results from improved knowledge of

— absorption of active substances at the site of administration, e. g. stomach, intestine, skin;

— distribution of active substances within the body;

— metabolism;

— elimination, in other words;

— the pharmacokinetics of active substances.

We all learned that there must be a minimum concentration (C_{min}) of the active substance at the site of action if a pharmacological effect is to be achieved. Since it is often difficult, if not impossible, to determine the concentration at the site of action, the minimum concentration in the plasma is usually measured instead. Toxicology shows us that a maximum concentration (C_{max}) in the blood plasma can be determined for every active substance. Above this concentration side effects develop or the ratio of effect to side effects becomes unacceptably low.

*) Lecture presented during the 32nd Annual Meeting of the Kolloid-Gesellschaft, Berlin October 2-4, 1985.

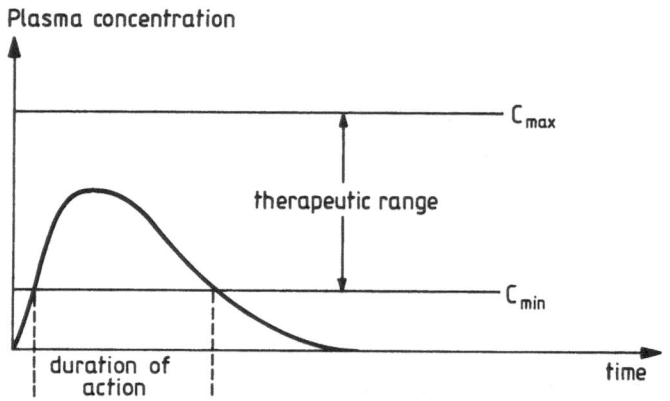

Fig. 1. Basic pharmacokinetic terms

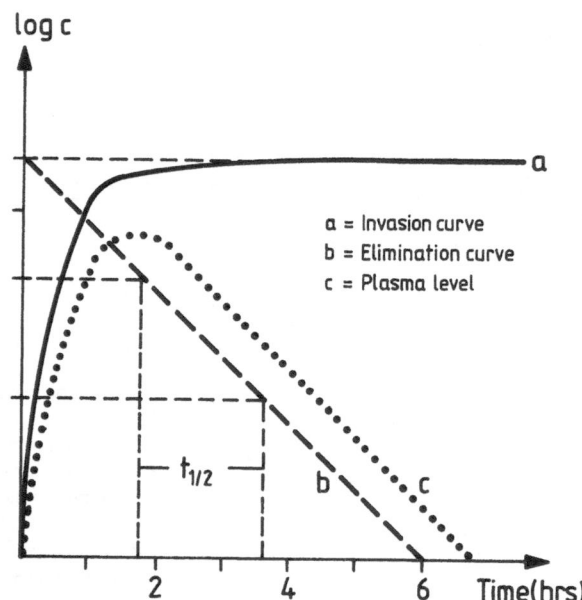

Fig. 3. Plasma level: two compartment model

The range of concentration remaining between these two values is the therapeutic range of the drug. The concentration measured in the plasma after administration of a certain amount of the substance is described as the plasma level (see Fig. 1).

Once the clinical paediatrician, Dost, from Gießen, had established this basic fact of pharmacokinetics (1953), it was very quickly discovered that different plasma levels could produce very different degrees of success in therapy. Or in other words: for many therapeutic goals, there are optimum plasma levels which can be determined. Therefore, in the case of an hypnotic, rapid onset is expected but, on the other hand, the concentration should be below the C_{min} value before the patient awakes. An analgesic should also

take effect quickly, but the effect should persist for as long as possible (see Fig. 2).

Analysis of the plasma levels shows that the curve results from the accumulation of several individual processes. The drug which enters the bloodstream is then eliminated according to kinetics specific to that drug (Fig. 3).

Before elimination can begin, however, the drug must be absorbed by the intestinal tract, for instance, or it must permeate the skin.

The invasion process — as long as the drug is present in a sufficient concentration and in a form suitable for invasion — is described by the drug-specific invasion constant. However, if the drug release from the dosage from is slower than invasion, the dosage form determines the build-up and course of the plasma level. By designing a suitable dosage form, the pharmacist can therefore add considerably to the efficacy and quality of a drug. Consequently, depot forms can be produced by increasing the amount of active substance per dosage form while controlling the release rate.

Release can often be controlled by colloid systems or polymers as will be explained in the light of the following two examples.

Drug release from a W/O system in vitro

In vitro release studies under well defined conditions allow for an understanding of the physics behind

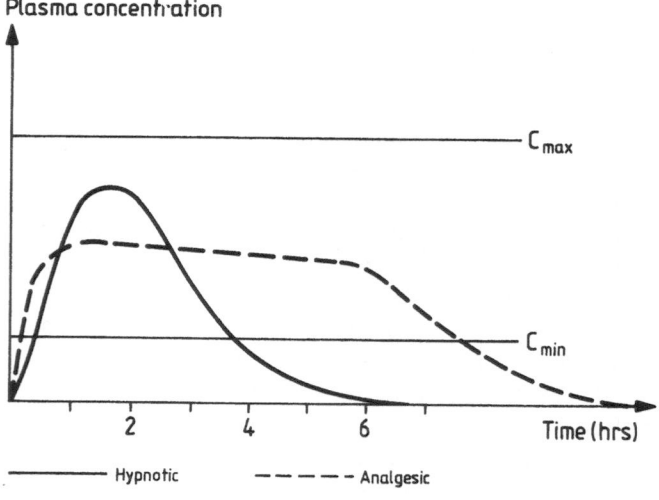

Fig. 2. Ideal plasma level curves

the process of drug release from emulsion systems. Such an understanding is an absolute prerequirement for any attempt at a meaningful study of percutaneous absorption.

W/O systems are emulsions in which the aqueous inner phase is dispersed as tiny drops into a coherent oily outer phase. This type of system is known in pharmacy as an ointment. In 1962 W. Higuchi [1] was the first to describe drug release from such a system. He established a square root relation between the amount substance $Q(t)$ released within a certain time.

$$Q(t) \sim \sqrt{t}$$

$$Q(t) = 2\,C_o F \sqrt{\frac{Dt}{\pi}}\,.$$

If we look at the structure of a W/O system on the one hand and the expression derived from Higuchi on the other, it immediately becomes clear that this equation can only correctly explain the facts to a certain extent since it explicitly regards the emulsion system as a continuum. If we give up this idea of a continuum, the following questions arise:

What is the function of the dispersed inner phase with respect to drug release?

What route does a drug with a given distribution coefficient choose during diffusion from the emulsion?

A simple diffusion cell is suitable for studying these problems in a W/O system. The emulsion system containing the substance is filled into one half as the donor phase, the other half is filled with a suitable acceptor medium. Water is a suitable acceptor medium for W/O systems.

If a very potent hydrophilic marker is worked into the dispersed aqueous inner phase, e. g. ethidium bromide, it can be established in a diffusion experiment that drug release from the emulsion system cannot be demonstrated with today's analytical methods.

The drug must obviously first move from the inner phase to the coherent outer phase which also forms the boundary surface for the acceptor medium. Only at this point is diffusion into the acceptor medium possible.

If varying quantities of the corticosteroid fluocortolon are incorporated into a W/O system[1]), this produces the release curves shown in Figure 4:

[1]) Neribas®-Salbe.

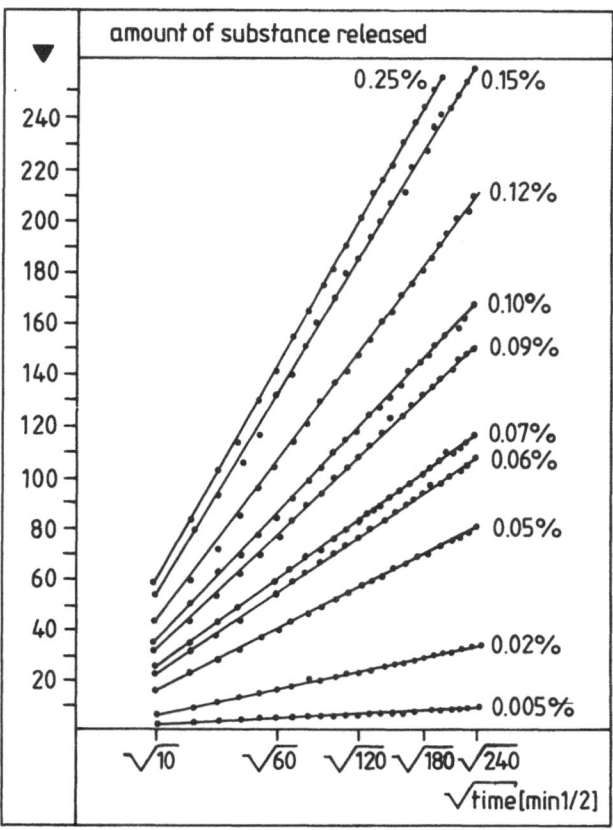

Fig. 4. Release of fluocortolon from W/O base

According to Fick's second law

$$\frac{\delta c}{\delta t} = D\,\frac{\delta^2 c}{\delta x^2}\,; \qquad \frac{\delta c}{\delta x} = \text{drug gradient}$$

the slope of the lines increases as the content of active substance increases (= drug slope becomes steeper).

Using the relation derived by Higuchi [2], the effective diffusion coefficients can be calculated from the slopes:

$$D = \left(\frac{\text{slope}}{2\,C_o \times F}\right) \cdot \frac{\pi}{60} \quad (\text{cm}^2/\text{sec})\,.$$

As can be seen in Table 1, they are virtually all identical up to a drug concentration of 0.12 %. As the last numerical value deviates significantly from the others,

Table 1. Release of fluocortolon from a W/O ointment

Ointment applied: 250 mg; thickness of layer: 0.5 mm

Conc. %	Content of active substance μg/cm³	Active substance released	Active substance released %	Diffusion coefficient cm²/sec.	Slope
0.005	12.5	7.28 μg	64 %	4.62×10^{-8}	0.54
0.02	50.0	30.3 μg	60 %	5.01×10^{-8}	2.13
0.05	125.0	71 μg	56.8 %	4.40×10^{-8}	5.17
0.06	150.0	97 μg	64.6 %	5.70×10^{-8}	6.97
0.07	175.0	102.9 μg	58.3 %	4.71×10^{-8}	7.54
0.075	187.5	131.3 μg	69.8 %	6.69×10^{-8}	9.71
0.09	225.0	142.1 μg	63 %	5.44×10^{-8}	9.88
0.10	250.0	157.8 μg	63.2 %	5.43×10^{-8}	10.83
0.12	300.0	186.0 μg	62 %	5.24×10^{-8}	13.48
0.15	375.0	222.7 μg	59.6 %	4.81×10^{-8}	16.93
0.25	625.0	237.1 μg	37.8 %	1.96×10^{-8}	18.28

$\bar{D} = 5.211 \times 10^{-8} \pm 0.67 \times 10^{-8}$ cm²/sec

it should not be included when calculating the mean value. The mean value effective coefficient is therefore

$$D = 5.21 \times 10^{-8} + 0.67 \times 10^{-8} \text{ (cm}^2/\text{sec).}$$

If we compare the slopes of the release curves with those for the drug concentrations of the preparations as shown in Figure 5, a linear relation emerges between these two parameters up to a concentration of $c = 0.15$ %. The gradient for $c = 0.25$ % no longer fits this correlation.

If we assume, as Higuchi did, that the total amount of drug is evenly distributed throughout the preparation, the diffusion coefficient should also be $D = 5.21 \times$

10^{-8} (cm²/sec.), even where the drug concentration is $c = 0.25$ %. The difference in the release pattern, i. e. the low slope, can be explained by the small amount of substance dissolved.

If we solve Higuchi's equation for the diffusion coefficient according to the initial concentration C_0 we obtain

$$C_o = \frac{Q(t)}{2F} \sqrt{\frac{\pi}{Dt}} \; .$$

This relation yields a freely diffusable, i. e. dissolved, quantity of active substance of 383.6 μg/cm³. With a 0.25 % preparation, this means only

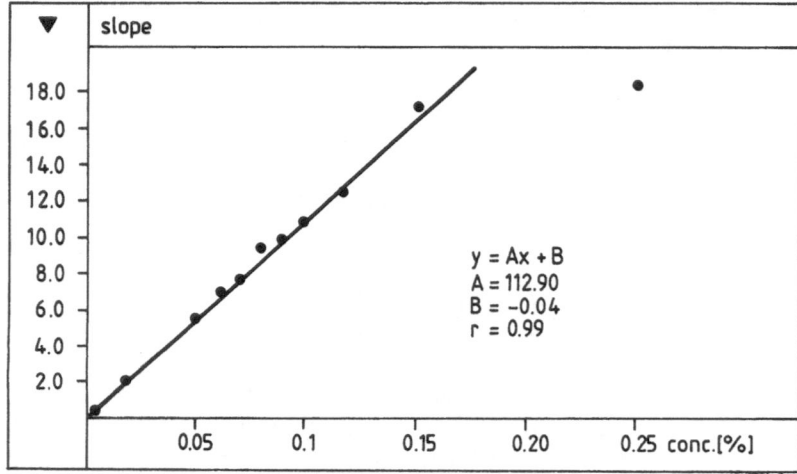

Fig. 5. Correlation between gradient and concentration of active substance in the preparations

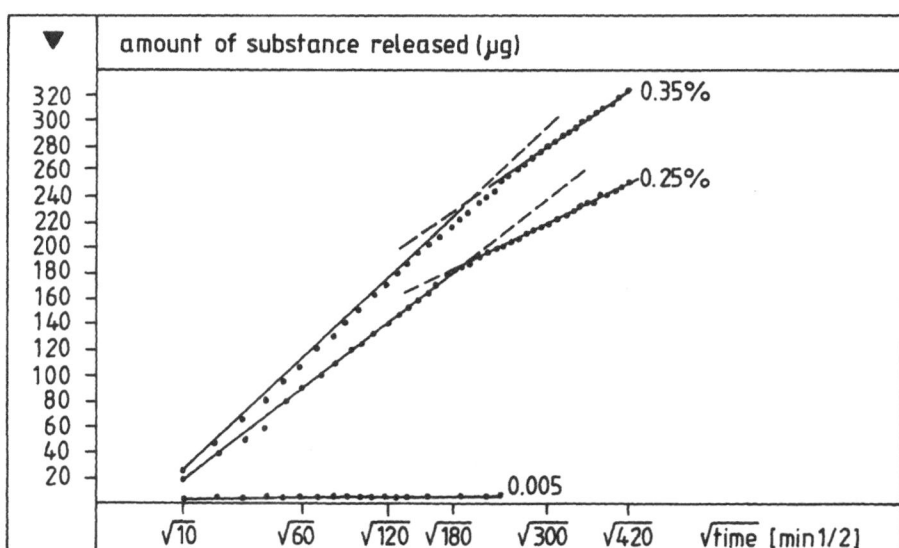

Fig. 6. Release of fluocortolon from a W/O base

about 61 % of the drug used is dissolved and freely diffusable.

The low diffusion coefficient observed at an active substance concentration of 0.15 % is not in fact different from the mean diffusion coefficient to a statistically significant degree.

If we take the mean diffusion coefficient, however, to determine how much of the drug has dissolved, it transpires that in this case, too, the solubility threshold has again been exceeded. Only 80 % of the drug used has actually dissolved.

The availability of a drug from a preparation is defined by the rate and extent of drug release from the preparation on the basis of the definition of bioavailability.

To determine the extent of drug release from the W/O emulsion, the thickness of the layer of ointment applied should be restricted to 0.20 mm. This makes it possible to monitor the total amount of substance released over a short period.

If the quantity of substance released is plotted over the square root of the time, the measured values no longer lie on one line, rather, two linear intercepts can be calculated (Fig. 6).

The first linear intercept lasts about 180 minutes for both preparations. Following a non-linear intercept lasting around 70 minutes, a second linear intercept can be observed.

This curve clearly contradicts the assumption that the active substance is evenly distributed in the base. The effective diffusion coefficients which can be calculated from the second linear intercept no longer give the drug release controlled by the emulsion's resistance of diffusion.

In the preparations used for dermal application we can usually identify the following components (= phases):
— water phase
— oil phase
— crystal phase.

The drug partially or wholly dissolves in the oil and water phase. After diffusion, the concentrations of active substance in both phases are in equilibrium according to Nernst. If there is still any drug in suspended form, this can begin dissolving in either the oil or water phase until the solubility threshold is reached.

Because of the distribution equilibrium between the oil and the water phases, it must be assumed for thermodynamic reasons that there is also equilibrium between both the water and crystal phases and the oil and crystal phases.

If this and the above findings on the release of highly water soluble substances are taken into consideration, the following model can be used for drug release from an emulsion system:

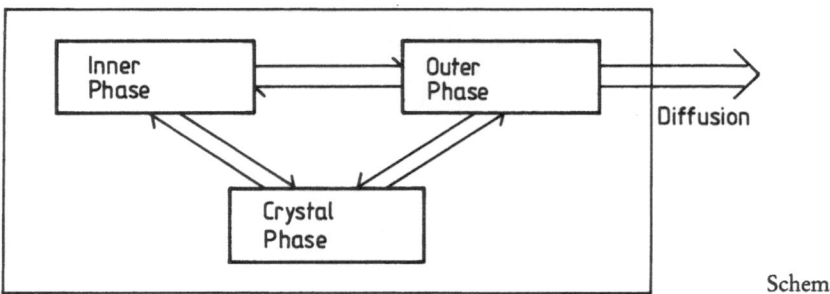

Schema 1

The crystal and dispersed water phase thus form the depot of active substance. The extent to which recourse is taken to this reserve during drug release depends on the ratio of the diffusion rate from the oil phase to the rate of equilibrium between water and oil phase and, hence, the dispersing capacity of the system.

The differential equations describing this system are explicitly no longer soluble. It is possible to study this problem using Markov chains. An analysis of this type — which cannot be described in detail here — allows the above findings to be simulated. Above all, they reveal the significance of the dispersed inner phase as a depot.

So far I have only discussed drop emulsions with a finely dispersed inner phase and a coherent outer phase. This type of system is characterised by a composition of 40–50 % water, about 5 % emulsifier and 40–50 % fat.

If the amount of emulsifier is increased to above the level required to cover the boundary surfaces between the water and oil phases, the emulsifier molecules can form other structures such as micellar aggregates in hexagonal or lamellar form. On the basis of packaging criteria, Israelachvili, Barber and Ninham specify the conditions for forming the various structures.

If we consider that migration as a prerequisite for diffusion is still possible in two directions in the lamellar phase, but only one direction is possible for migration in the hexagonal phase, it becomes clear that emulsions with a high percentage of liquid crystallins have a markedly lower diffusion coefficient than drop emulsions.

Drug release from a polymer system

Following the use of silicone rubber in implants over a considerable period of time, antibiotics were first incorporated in silicone polymer depots in the 1960s.

The aim was to achieve constant drug release over as long a period as possible. It transpired, however, that constant drug release was only possible if a silicone tube had been filled with a suspension of the drug. This guaranteed that the concentration of disolved active substance on the inner side of the silicone tube remained constant at $C = C_s$. The concentration gradient over the membrane thus also remained constant (see Fig. 7).

In the interest of safety this type of dosage form has only been used experimentally. Only in the case of so-called monolithic depot forms can uncontrolled drug release due to damage to the drug reservoir definitely be prevented.

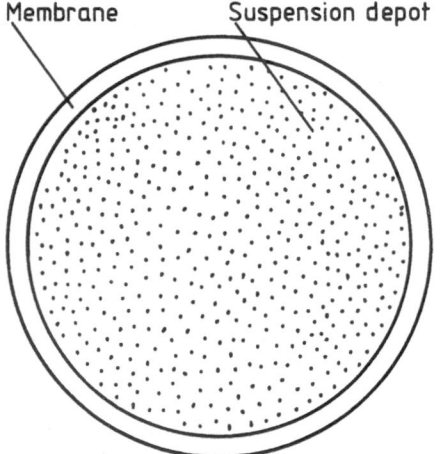

Concentration at the boundary surface
Depot/membrane
$C = C_s$ = constant
C_s = solubility in suspension agent

Fig. 7. Depot dosage form: suspension depot

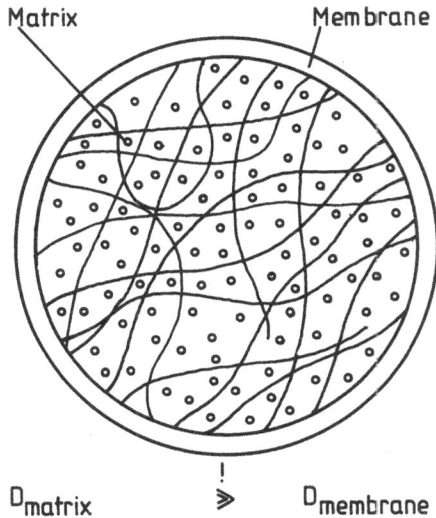

Fig. 8. Depot dosage form: matrix depot

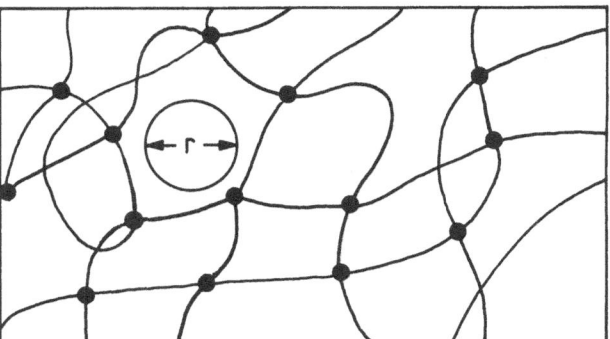

Fig. 9. Diffusion through rubber-elastic network

In this case the active substance is in a finely distributed partially dissolved form in a fixed polymer matrix. During drug release the extent of diffusion increases. Consequently we observe a kinetic decrease in the rate of release to the order of 0.5.

It would be possible to produce an implantable depot with a constant release rate over a lengthy period of time on one hand, but with no risk of uncontrolled drug release on the other, if a matrix depot (Fig. 8) with a very high diffusion coefficient is covered by a membrane with a very much lower diffusion coefficient.

This type of programme essentially presupposes that it is possible to set reproducible diffusion coefficients for polymer systems of this kind.

At molecular level, diffusion of a small molecule within a three-dimensional rubber elastic polymer network can be treated according to the theory of free volume developed by Cohen and Turnbull [3] for solid elasticity and applied by Vrentas and Duda [4] in transport processes.

This theory posits a substance molecule dissolved in a polymer inside a cage composed of polymer chains (Fig. 9). These perform thermic oscillations. There are collisions with the substance molecule which result in an impulse exchange. If on collision an impulse is exchanged strong enough to overcome the electric interactions with the polymer environment, the molecule can migrate on condition that at the same time it finds a neighbouring hollow, free volume, large enough to accommodate the substance molecule. Both

events, impulse exchange and the finding of a sufficiently large free volume are independent of each other. Consequently, this results in the probability of such migration as the product of the probabilities $f(p)$ and $g(v)$ of sufficiently strong impulse exchange or of the presence of a suitable free volume. The diffusion coefficient can be given approximately as

$$D \approx f(g) \cdot g(v).$$

As long as the same polymer system and the same substance are being considered, the first term $f(g)$ can be regarded as a constant.

It therefore follows that

$$D \approx K \cdot g(v).$$

That means the diffusion coefficient is determined by the free volume and therefore the network structure.

Of the various polydimethyl siloxanes, the following are available:
— linear vinyl-terminated polydimethyl siloxanes
— linear polydimethyl siloxane with Si-H-groups at the terminal points, with a few branches per chain
— cyclic tetramethyl-tetravinyl-cyclotetrasiloxane

By mixing these components we can build up networks of differing chain length between the connecting points. The connection between the composition of the mixture and the length of network lines which can be achieved can be shown by means of a polynomial. Figure 10 shows how well the network line lengths obtained experimentally agree with those used in the polynomial to calculate the formula.

By combining polymer rubbers of varying network thickness which satisfy the above condition $D_{\text{Matr.}} \ll$

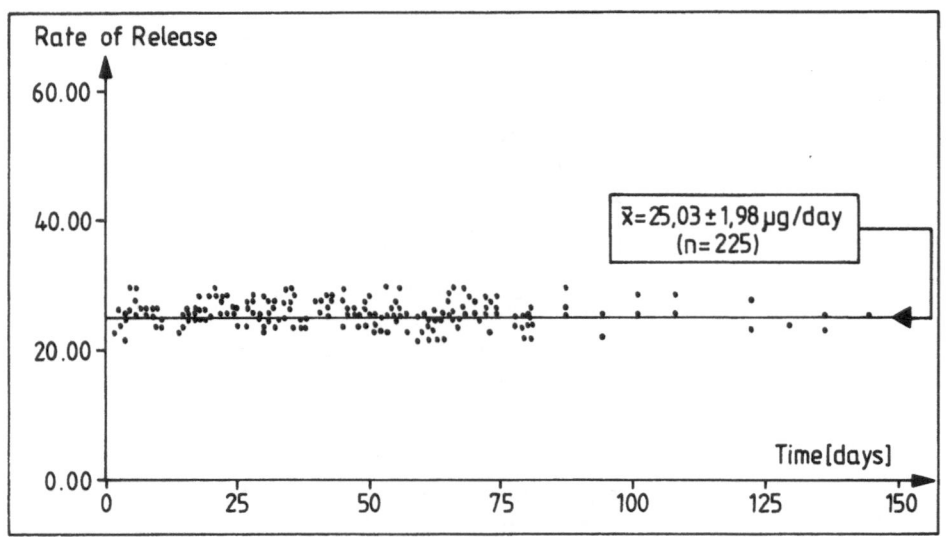

Fig. 10. Release of a progestin

$D_{Membr.}$, matrix depot systems can be produced which allow constant release rates over a long period of time. In Figure 11 we see the in vitro release of a progestin from a depot system over more than 150 days. As you see, the release rates remain extremely constant.

I have presented two systems which can be placed at either extreme of our main subject.

I hope I have been able to illustrate how pharmacy, especially today, has gained from findings in the field of colloid systems and polymers. In view of the increasing significance of specific and/or controlled drug release more and more value will be attached to cooperation between colloid and polymer chemists as well as physicists and pharmacists.

Table 3. Comparison of networks used to calculate formula and the values obtained experimentally

Given network line length to calculate formula in (Si-O)-units	Network line length obtained experimentally in (Si-O)-units	
24.	19.69	
	25.54	
	25.59	
	24.63	
	25.90	$\bar{x} = 24.27$
27.	27.90	
	26.78	
	27.01	
	26.78	
	26.52	$\bar{x} = 26.80$
49.	49.29	
	48.91	
	47.59	
	48.89	
	48.38	$\bar{x} = 48.55$
160.	160.75	
	159.92	
	159.81	
	159.41	
	160.55	$\bar{x} = 160.05$

Table 2. Release of fluocortolon from a W/O ointment

Ointment applied: 100 mg: Thickness of layer: 0.20 mm; volumes: 0.40 cm³

Concentration	0.25 %	0.35 %
Content of active substance/cm³	625 µg	875 µg
Quantity of active substance released within 180 minutes	177.7 µg	219.2 µg
Quantity of active substance released within 180 minutes in %	50.8 %	62.6 %
Quantity of active substance released within 7 hours	236.1 µg	317.4 µg
Quantity of active substance released within 7 hours in %	94.4 %	90.7 %
Diffusion coefficient of the first linear segment	1.47×10^{-8} cm²/sec	1.14×10^{-8} cm²/sec
Freely diffusible quantity of active substance	331.9 µg/cm³	409.5 µg/cm³

References

1. Higuchi WJ (1962) J Pharm Sci 51:802
2. Koizumi T, Higuchi WJ (1968) J Pharm Sci 57:87
3. Cohen MH, Turnbull D (1959) J Chem Phys 31:1164–1169; Turnbull D, Cohen MH (1961) J Chem Phys 34:120–125
4. Vrentas JS, Duda JL (1976) Macromolecules 9:785–790; Vrentas JS, Duda JL (1977) J Polym Sci, Polym Phys Ed 15:403–416; Vrentas JS, Duda JL (1977) J Polym Sci, Polym Phys Ed 15:417–439

Received December 5, 1985; accepted February 28, 1986

Author's address:

Dr. I. Zimmermann
Boehringer Ingelheim Zentrale GmbH
FL Pharm. Forschung und Entwicklung
D-6507 Ingelheim am Rhein, F.R.G.

Progress in Colloid & Polymer Science   Progr Colloid & Polymer Sci 72:28–36 (1986)

Coagulation of latex dispersions by inorganic salts: structural effects*)

R. Zimehl and G. Lagaly

Institut für anorganische Chemie der Universität Kiel, F.R.G.

Abstract: Three types of latices were prepared by polymerization of styrene. In type 1 (without emulsifier) and type 2 series (with sodium dodecyl sulfate) the polymerization was initiated by different amounts of peroxodisulfate. Type 3 latices are co-polymers of styrene and styrene sulfonate prepared without initiator. The dispersions reveal significant differences when coagulated with salts. The differences are related to different surface structures. Type 1 latices are considered "classical" latices with smooth surface which are stabilized by electrostatic repulsion. Latex particles of type 3 are assumed to be surrounded by an envelope or a corona of hairs. The hairs add a considerable amount of steric stabilization to the electrostatic stabilization.

In both series the coagulation power increases from Li^+ to Cs^+ which is explained by changes of the water structure. For the same reason, the coagulation concentrations are also sensitive to the kind of co-ions (anions).

Key words: Colloids, coagulation, latex, salt effects, steric stabilization, water structure.

Introduction

Aqueous latex dispersions are commonly considered excellent models of electrostatically stabilized colloidal dispersions. Addition of inorganic salts causes the dispersion to coagulate. The critical coagulation concentrations c_K and their dependence on the gegen ion valency can follow the DLVO theory, but, in point of fact, not every latex dispersion behaves in this way; exceptional high or low coagulation concentrations can be observed. The differences often become strikingly evident when the dispersions are coagulated by inorganic salts in the presence of organic materials [1].

Latex dispersions produced industrially for practical applications generally contain several compounds admixed which could be the cause of the differences. However, even monodisperse latex dispersions which were carefully purified can reveal striking differences.

A decisive cause is seen in the surface structure of the latex particles. The particles are usually regarded as isolated spheres with distinct surface charge densities or surface potentials. The surface, however, may be smooth in one case, and, in another case, may possess "hairs" radiating away from surface [2–5]. It is easy to understand that charged macromolecules formed during the polymerization shoot out of the surface. The macromolecules may return to the surface as loops or may radiate into the solution as tails. They may also be more or less coiled to form envelopes around the particles which can change their thickness with salt concentration and solvent parameters. When an envelope or a corona of hairs surrounds the particles, a steric component contributes to stabilization (Fig. 1). Increasing steric stabilization should reduce the sensitivity of the dispersed particles toward salts.

The different behavior of latex dispersions will be exemplified with three series of polystyrene latex dispersions (Table 1):

Type 1 latices were prepared by polymerization of styrene in the absence of emulsifiers but with different concentrations of potassium peroxodisulfate ("KPS") as polymerization initiator.

Type 2 latices are polystyrene latices prepared in the presence of constant amounts of sodium dodecyl sulfate ("SDS") and varying amounts of KPS.

---

*) Lecture presented during the 32nd Annual Meeting of the Kolloid-Gesellschaft, Berlin October 2–4, 1985.

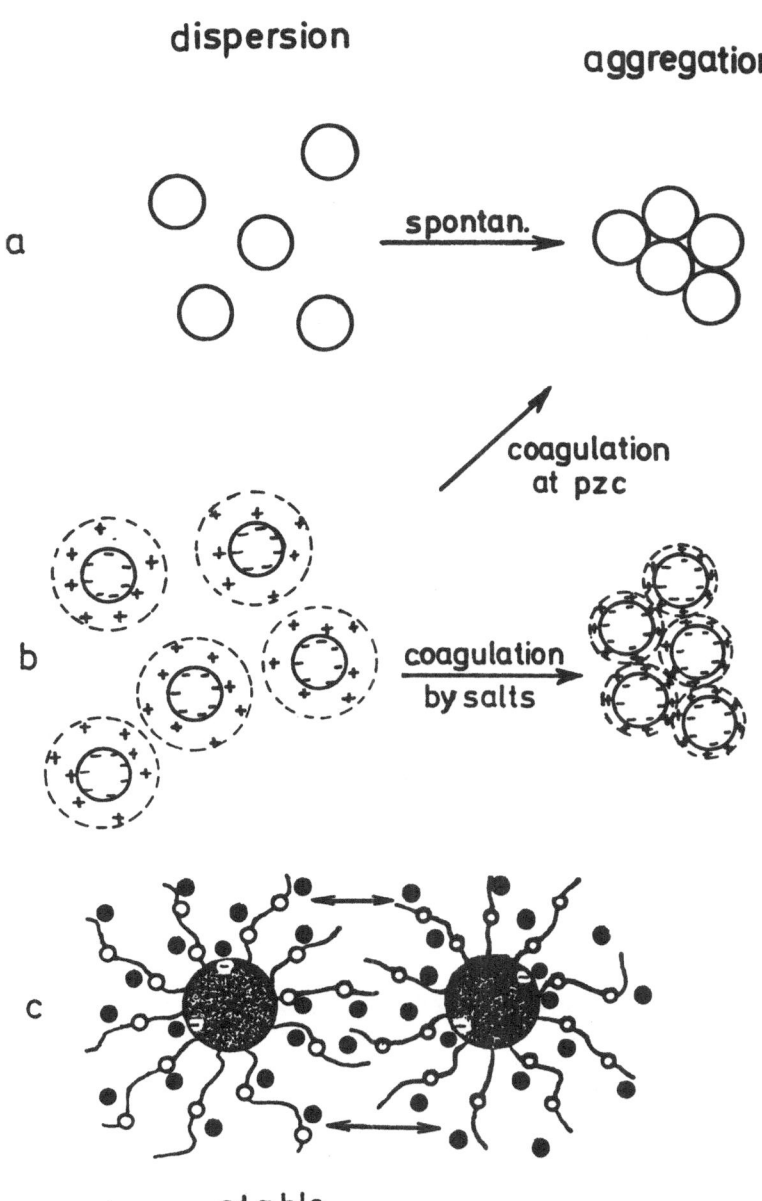

Fig. 1. Stabilization of latex dispersions; a) unstable dispersions; b) electrostatic stabilization; c) electrosteric stabilization

Type 3 latices are co-polymers of styrene and potassium styrene sulfonate ("KSS") in different ratios.

Experimental

Preparation of type 1 latices

The latices of series PS (Table 1) were prepared from 25 ml of styrene dispersed in 180 ml of twice – distilled water (1-L round-bottom four-necked flask with condenser, nitrogen inlet, PTFE paddle stirrer and contact thermometer). Under vigorous stirring and intense nitrogen gas flow the temperature was raised until refluxing. The nitrogen gas flow was reduced, and the desired amount of potassium peroxodisulfate dissolved in 25 ml of twice – distilled water was added and washed in with further 20 ml of water. The stirring frequency was then reduced to about 50 rpm. The reaction started very quickly and was held at 100 °C for 6 hours. The dispersion was filtered first over glass wool, then through a paper filter and, if necessary, dialyzed [6–8]. (Latex PS 3 revealed a peculiar behavior among the PS series. The undialysed dispersion became unstable after several weeks. The dispersion separated in an upper phase containing large flocs and a lower phase with the appearance of a very diluted latex dispersion).

Latex 100/4/2 (mean particle diameter: 90 nm) was prepared under the same conditions from 2 ml of styrene in a total of 800 ml of water (instead of 25 ml of styrene in 225 ml of water) exactly follow-

Progress in Colloid & Polymer Science, Vol. 72 (1986)

Table 1. Preparation of PS-latex dispersions at 100 °C. Different amounts of KPS, SDS or KSS were added to 25 ml styrene dispersed in a total of 225 ml of water. KPS = potassium peroxodisulfate, SDS = sodium dodecylsulfate, KSS = potassium styrene sulfonate (4-vinyl benzene sulfonate)

Sample	g KPS	g SDS	g KSS	$\bar{d}$[a]) (nm)	pH before dialysis	after dialysis
PS 1	0.01	—	—	240	3.8	5.7
PS 2	0.05	—	—	250	2.8	5.9
PS 3	0.10	—	—	(600)	2.5	4.1
PS 4	0.25	—	—	330	2.1	5.5
PS 5	1.00	—	—	530	1.5	4.1
PS 6	2.00	—	—	490	1.3	4.7
SDS 7	—	0.25	—	1060	3.6	4.3
SDS 12	0.10	0.25	—	560	2.8	
SDS 11	0.25	0.25	—	850	2.1	
SDS 5	0.50	0.25	—	250	1.9	3.7
KSS 0	1.00	—	—	850[b])	2.2	5.1
KSS 1	0.50	—	1.00	80[b])	2.4	3.2
KSS 9	—	—	0.05	570[c])	4.5	5.9
KSS 8	—	—	0.10	330[c])	4.9	5.9
KSS 7	—	—	0.50	190[d])	4.5	5.1
KSS 4	—	—	1.00	120[b])	4.1	3.9
KSS 3	—	—	2.00	80[b])	7.3	3.4
KSS 5	—	—	3.00	80[b])	7.1	3.5
KSS 6	—	—	4.00	500	7.5	3.5
KSS 10	—	—	4.50	440	7.1	3.6
KSS 2	—	—	5.00	290	8.2	3.4

[a]) by sedimentation; [b]) in 90 % methanol/H_2O; [c]) in 60 % methanol/H_2O; [d]) in 77 % methanol/H_2O

ing the procedure of Goodwin et al. [9]. The particle diameters of this latex and other samples (not shown in Table 1) prepared under varying conditions (ionic strength, initiator and monomer concentration, temperature) follow the relation between particle diameter and experimental parameters established by Goodwin et al. [9].

Preparation of type 2 latices

These latices were prepared from 25 ml of styrene and 0.25 g of sodium dodecyl sulfate ("SDS") in 180 ml of twice – distilled water using the same equipment as described before. After reaching a constant reflux, the initiator (KPS dissolved in 25 ml of twice – distilled water), was added, washed in with 20 ml of water, and the emulsion was allowed to react for 8 h at 100 °C. In the case of latex SDS 7 (prepared without KPS) the reaction needed several minutes to start. In all other cases, the reaction started almost immediately. The latex dispersions were used without dialysis. Latex SDS 5 produced a pronounced violet iridescence during preparation and handling.

Preparation of type 3 latices

The procedure was similar to the other preparations; the desired quantity of potassium 4-vinyl benzene sulfonate (potassium styrene sulfonate "KSS") was dissolved in 25 ml of water and added to the styrene/water emulsion instead of the KPS solution.

The initiation time decreased from 10–15 min at the lowest KSS content to 2–5 min at the highest KSS/ styrene ratios (samples KSS 10 and KSS 2).

The latices KSS 9, 4 and 6, 10, 2 are not strongly monodisperse but the proportion of large particles (aggregates?) is modest. The mean particle diameter proceeds through a pronounced minimum at ratios of 1–3 g KSS/25 ml styrene.

Characterization of the latices

The mean particle diameters (Table 1) were determined by a simple light scattering method, dynamic light scattering (photon correlation spectroscopy [10] and AWPS [11]) and by sedimentation in the centrifugal field. In the simplified procedure described by van den Esker and Pieper [12] the mean particle size is derived from turbidity measurements of the very diluted latex dispersions. A commercial particle size distribution analyser (CAPA-500, Horiba Europe GmbH, 6374 Steinheim, F.R.G.) was used for particle size determinations by sedimentation. Several samples were examined by transmission and scanning electron microscopy.

The surface charge density was measured by conductometric titration of the dispersions with 0.01 M KOH after exchanging the gegen ions and co-ions with a mixed-bed ion exchanger (Lewatit CNP 80 + MP 62) [13,14].

Coagulation experiments

The dispersions were diluted with twice – distilled water to give a latex volume fraction of $4 \cdot 10^{-4}$; the dispersions had then pH ≈ 6. The critical coagulation concentrations c_K were measured as soon as possible after dialysis. In particular, the latex dispersions prepared with KPS as initiator can change their properties after standing for a longer time. The sulfate ester groups are hydrolyzed which reduces the surface charge density. The remaining —OH groups render the surface more hydrophilic [15].

Test-tube experiments [16]:

A series of dispersions containing increasing amounts of salts were prepared by adding 1 ml of the salt solution to 1 ml of the latex dispersion (final latex volume fraction: $2 \cdot 10^{-4}$). After shaking, the dispersions were allowed to stand for 24 h. It was then decided which salt concentrations cause coagulation. When required, additional series were prepared by increasing the salt concentration in smaller steps. This simple method should not be considered obtuse because of the lack of electronic equipment. It allows a critical and reliable examination of the stability and gives information about the type of aggregates formed.

Photometric measurements:

In addition, the critical coagulations concentrations c_K were determined by photometric turbidity measurements [8, 13, 17–20]. The optical density $D = \lg I_0/I$ (I_0 = intensity before, I = intensity during coagulation) is recorded as a function of time t at the early states of coagulation (up to some minutes). From dD/dt of the linear section of the curves the stability ratio and c_K are derived. The linear section is often short and dD/dt is difficult to measure exactly. In series of experiments we preferred a procedure similar to the method described by Bibeau and Matijević [21]. Relative turbidities τ_0/τ were plotted against the concentration (τ_0, τ: turbidities of the pure latex dispersion and of the dispersion 1 min after addition of salt). The shape of the curves is similar to the $\log W = f(c)$ curves. The concentration above which τ_0/τ remains constant (or, as often observed, increases slightly), is taken as the critical coagulation concentration. As long as D increases linearly with time, this procedure yields the same c_K as the $\log W = f(c)$ curves.

Generally, the curves $D = f(t)$ deviated from linearity somewhat above $t = 1$ min. A significant deviation was observed in several cases when the dispersion was coagulated by salts in the presence of organic compounds [1]. The deviations are sensitive to the type of the organic compounds, and the impression is produced that they indicate different aggregation mechanisms.

Results

Type 1 latices:

The latex dispersions of this series were prepared by emulsion polymerization of styrene without emulsifier but with increasing amounts of KPS as

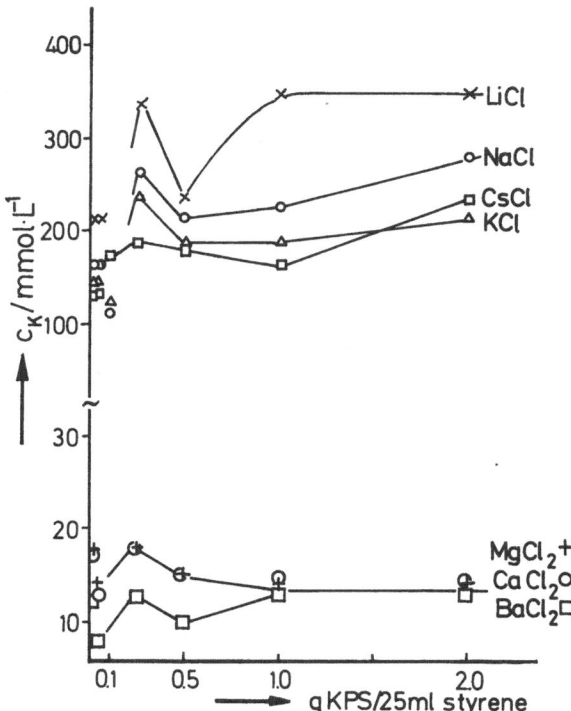

Fig. 2. Critical coagulation concentrations of type 1 latices, prepared in the presence of different amounts of potassium peroxodisulfate

initiator.[1]) With increasing KPS/styrene ratio, the surface charge density should increase because the charges are produced by sulfate groups (C–$OSO_3^{\ominus}$). In fact, the surface charge density σ_o (obtained from conductometric titrations) increases with the ratio KPS/styrene from about 0.2 μCcm^{-2} to about 20 μCcm^{-2}.

The coagulation concentrations of the alkali chlorides are collected in Figure 2. Above 0.5 g KPS/25 ml styrene c_K increases with this ratio. The increase is most pronounced for LiCl. CsCl exhibits a slight minimum. In nearly all cases, the coagulation power increases from LiCl to CsCl. Below 0.5 g KPS/25 ml styrene maxima and minima are observed and the order of the salts changes below 0.25 g KPS/25 ml styrene. In the presence of the divalent cations Mg^{2+}, Ca^{2+} and Ba^{2+} c_K becomes independent of the ratio KPS/styrene and the kind of the cation (above 0.5 g KPS/25 ml styrene).

---

[1]) In all Figures and Tables the ratio KPS/styrene (or KSS/styrene) gives the ratio of the compounds as added to the preparations. This ratio does not necessarily correspond to the ratio of sulfate groups/styrene segment.

Table 2: Critical coagulation concentrations c_K of type 2 latices (test-tube tests)

Latex	SDS 7	SDS 12	SDS 11	SDS 5
g KPS/25 ml styrene	0	0.10	0.25	0.50
	critical coagulation concentration (mmol · L^{-1})			
LiCl	> 1800	120	100	375
NaCl	188	119	178	376
KCl	263	138	138	188
CsCl	> 900	75	63	200
TMACl[a]	2985	71	60	188
GuHCl[b]	299	89	79	158
MgCl$_2$	7.5	20	25	25
CaCl$_2$	6.3	13	11.3	12.5
BaCl$_2$	3.5	10	11.3	11.3
LaCl$_3$	0.25	0.25	0.25	0.25
C$_{12}$PyCl[c]	0.10	0.004	0.006	0.035

[a]) tetramethylammonium chloride; [b]) guanidinium hydrochloride; [c]) n-dodecylpyridinium chloride

Type 2 latices:

The latices were prepared by adding increasing amounts of KPS to styrene in the presence of a *constant amount* of sodium dodecylsulfate. As indicated in Table 2 the c_K of alkali chlorides do not vary systematically either with increasing ratio KPS/styrene, or with the type of the cation. Exceptionally high c_K are observed for LiCl and CsCl and latex SDS 7. The c_K of the divalent cations are nearly independent of the ratio KPS/styrene; somewhat lower values are observed for SDS 7. The coagulation concentration of LaCl$_3$ is constant for all preparations. This confirms the viewpoints [21–23] that the charge of the particles mainly arises from adsorbed surfactant ions.

Table 2 includes the coagulation concentrations of some organic salts. Tetramethylammonium chloride behaves like CsCl in all KPS containing systems but exhibits a very weak coagulation power towards latex SDS 7. Guanidinium chloride shows the same trend. Dodecylpyridinium chloride as cationic surfactant coagulates the latex dispersions at very low concentrations. An about twenty-fold concentration is required to coagulate SDS 7.

Type 3 latices:

The critical coagulation concentrations change in a characteristic way with the ratio KSS/styrene (Fig. 3).

The cations follow the order:

$$c_K: \text{Li}^+ > \text{Na}^+ > \text{K}^+ > \text{guanidinium} > \text{Cs}^+ > \text{Mg}^{2+}$$

$$> \text{Ca}^{2+}.$$

The critical concentration increases sharply above 3 g KSS/25 ml styrene reaching a distinct maximum at 4.5 g KSS/25 ml styrene. Particular attention should be drawn to the exceptional high salt stability of this latex dispersion: the coagulation concentrations of LiCl and NaCl are 2600 and 1760 mmol · L^{-1}, respectively.

The same trend is observed for Mg^{2+} and Ca^{2+} cations with a significant high stability of the latex KSS 10 toward Mg^{2+} ions. Ba^{2+} ions behave distinctly differently; the coagulation concentration increases sharply at very low KSS contents.

Anion effects

The critical coagulation concentrations are not only sensitive to the cations as gegen ions but depend also on the kind of the anion. Table 3 shows this effect for a type 1 and a type 3 latex. The coagulation concentrations for the type 1 latex increase from F$^-$ to SCN$^-$.

Table 3. Critical coagulation of different salts towards a type 1 latex (100/4/2) and a type 3 latex (KSS 2) (concentrations in mmol · L^{-1})

Salt	Latex 100/4/2[a]	Latex KSS 2[b]
LiCl	531	1900
NaCl	473	910
KCl	363	850
NH$_4$Cl	299	
RbCl	272	
CsCl	299	800
MgCl$_2$	47	15
CaCl$_2$	34	7.5
BaCl$_2$	24	12
LaCl$_3$	0.42	0.068
NaF	457	
NaCl	473	910
NaI	509	800
NaN$_3$	537	
NaSCN	562	800
NaClO$_4$		940
Na$_2$SO$_4$	275	250

[a]) photometrical determinations and test-tube tests; [b]) test-tube tests

The type 2 latex is coagulated by about the same concentration of sodium sulfate whereas the concentrations of the sodium salts of monovalent anions are higher and follow a different order:

$$SO_4^{2-} \lll I^- \approx SCN^- < Cl^- < ClO_4^-.$$

Discussion

Electrostatically stabilized dispersions

The type 1 dispersions (above 0.5 g KPS/25 ml styrene) contain latex particles in "classical" sense: spherical particles with smooth surfaces, without "hairs", and negatively charged by sulfate groups. The c_K of LiCl and MgCl$_2$, CaCl$_2$ indicate Stern potentials ψ_δ of 25–30 mV ($z = 1$–1.2 for monovalent gegen ions). The increase of c_K with increasing KPS/styrene ratio corresponds to a modest increase of the surface potential ψ_δ from 25 to 30 mV, because the c_K are very sensitive to ψ_δ at low potentials (Fig. 4) [19, 24]:

$$c_K = \frac{49.63 \, (4\pi\varepsilon\varepsilon_o)^3 (RT)^5}{F^6} \frac{\gamma^4}{A^2 v^6} \; [\text{mol} \cdot \text{m}^{-3}]$$

$$= 3.84 \times 10^{-39} \, \gamma^4/A^2 v^6 \; [\text{mol} \cdot \text{L}^{-1}] \qquad (1)$$

with: $\gamma = (e^{z/2} - 1)/(e^{z/2} + 1)$ and $z = ve_o \, \psi_\delta/RT$. For particles distinctly smaller than 10^3 nm, corrections have to be introduced [24] which give somewhat higher values of c_K.

The titratable surface charge density corresponds to higher surface potentials than derived from the coagulation concentrations, indicating a considerable part of the gegen ions to be bound in the Stern layer [14]. On this basis, it is not astonishing that the c_K of MgCl$_2$ and CaCl$_2$ decreases slightly with increasing ratio KPS/styrene (Fig. 2). The higher charge density increases the proportion of the divalent gegen ions in the Stern layer and reduces ψ_δ which determines the stability toward salts. The Stern layer adsorption of the multivalent gegen ions is also evident from the c_K of LaCl$_3$ which are generally lower than calculated by Equation [1].

At low KPS/styrene ratios the surface charge density is distinctly reduced so that (i) small changes of ψ_δ can produce considerable changes of c_K, and (ii) the stability can be mainly determined by factors other than electrostatic charges. Two main factors must be mentioned: The particle may behave as a real hydrophobic moiety and is then surrounded by water shells that are organized like water shells around apolar com-

pounds (as discussed in relation with hydrophobic interactions). An other possibility is that the particles carry an envelope or corona of macromolecules. At the present time, we cannot decide between both possibilities.

Influence of styrene sulfonate

Co-polymerization of styrene with styrene sulfonate ("KSS") should promote a "hairy" surface structure. During the polymerization charged groups are pushed off from the core to the surface, and the probability increases that ionic macromolecules protrude out of the surface. (The mean particle diameter seems to change with the latex concentration and solvent properties e. g. latex KSS 6 $\bar{d} = 500$ nm in H$_2$O, 370 nm in 90% methanol. It supports the idea of a corona of hairs or a "gel-like" envelope of macromolecules around a "hard" core. This needs further studies.)

The sharp increase of c_K above 3 g KSS/25 ml styrene is taken as an evidence of "hairs". At the present time, a more direct proof of this assumption is difficult to work out. When ionic macromolecules radiate away from the surface, the electrostatic stabilization is supported by steric stabilization (Fig. 1) [25]. The electrosteric stabilization decreases the sensitivity toward salts and increases c_K. In Figure 3 this effect is illustrated by a sharp increase of the coagulation concentrations of alkali chlorides at high ratios KSS/styrene. The stabilization effect can be suppressed by multivalent cations. The maximum of c_K still persists for MgCl$_2$ and is also detectable for CaCl$_2$ and latex KSS 10. In all other cases, the c_K of the divalent cations and La^{3+} are not significant different from the c_K of the type 1 and type 2 latex dispersions. Evidently, the di- and tri-valent cations can counteract the steric stabilization. One may imagine in a simple picture that the corona of macromolecular chains around the particles is contracted by interactions with the multivalent cations.

The low c_K of multivalent salts not corresponding to the high c_K of LiCl and NaCl, may be considered as a further evidence that stabilization is not purely electrostatic. When 2000 mmol $\cdot$ L^{-1} of monovalent gegen ions are required to coagulate the system, electrostatic stabilization demands coagulation concentrations of di- and tri-valent gegen ions of about 200 and 40 mmol $\cdot$ L^{-1} (or somewhat lower because of increased Stern adsorption). Observed are significant lower values of 7–30 (60) and 0.1 mmol $\cdot$ L^{-1}, respectively.

The astonishing increase of c_K for BaCl$_2$ and latices KSS 8 and 9 is inexplicable at the present time.

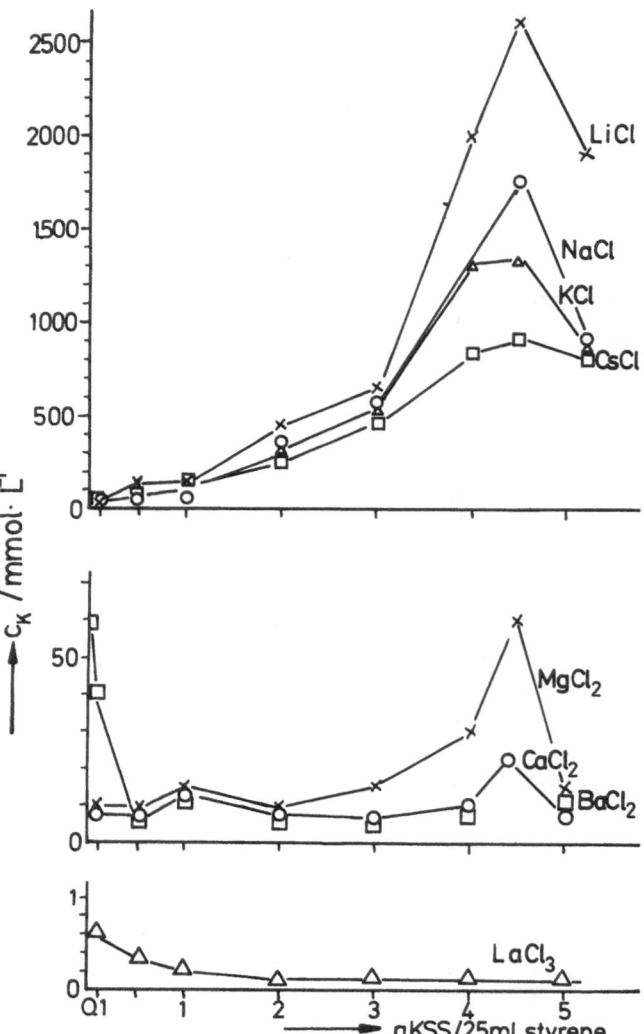

Fig. 3. Critical coagulation concentrations of type 3 latices, prepared in the presence of different amounts of potassium styrene sulfonate ("KSS")

The presence of surfactant anions on the surface and in the solution renders the behavior of these latex dispersions particularly complex. The surfactant ions are adsorbed on the surface of the particles or even may be deeply anchored with their tails in the latex moiety [23]. The coagulation power of the salts no longer follows the order observed for the other latex dispersions. The c_K of LiCl, CsCl and tetramethylammonium and SDS 7 dispersions are exceptional high. Since this latex has been prepared without initiator, the surface charge is very low and steric stabilization may be dominant. However, the analogous behavior of

LiCl and CsCl seems to be difficult to explain, particularly with regard to the following discussion.

Water structure effects

The interpretation of specific gegen ion effects exerted by alkali and alkaline-earth cations "on purely energetic grounds does not seem very feasible" [26]. An important contribution of entropic effects may arise from the influence of the ions on the water structure. The sequence of c_K in the order

$$Li^+ > Na^+ > K^+ > Cs^+ \text{ and } Mg^{2+} > Ca^{2+} > Ba^{2+}$$

is well-known as the Hofmeister series. The position of the cations within this series is related to their influence onto the water structure. From Li^+ to Cs^+ the breaking power increases. This is better expressed by the structure temperature T_{str} proposed by Bernal and Fowler [27] and studied in detail by Luck [28, 29]. The structure temperature is derived from the overtone bands of the water molecules in the electrolyte solutions. It is defined as the temperature of pure water with a hydrogen bond state similar to the electrolyte solution under consideration. The somewhat vague definitions "structure making" and "structure breaking" can then be replaced by a more tangible definition: ions with $T_{str} > T$ are structure breaking ions, ions with $T_{str} < T$ are structure making ions.

Increasing structure temperature ($Li^+ \rightarrow Cs^+$) reduces the hydration of the cations, and the probability is increased to find them as gegen ions near the particle surface. When the gegen ion concentration in the Stern layer increases, ψ_δ and the electrostatic repulsion between the particles are reduced: increasing structure temperature decreases the critical coagulation concentration. The model is basically similar to the model of hydration barriers as proposed by Healy et al. [30] and Ruckenstein [31].

Changes of the water structure also affect the electrosteric stabilization. Thus, the same order of the gegen ions is maintained. The effect is even somewhat intensified by the interaction between the chains and the water structure.

Interaction of surfactants with the water structure is a well-known phenomenon which leads to the association of the surfactants above a threshold concentration. It is thus understandable that in the presence of surfactant anions deviations from the Hofmeister series of the cations do occur. The mutual influence

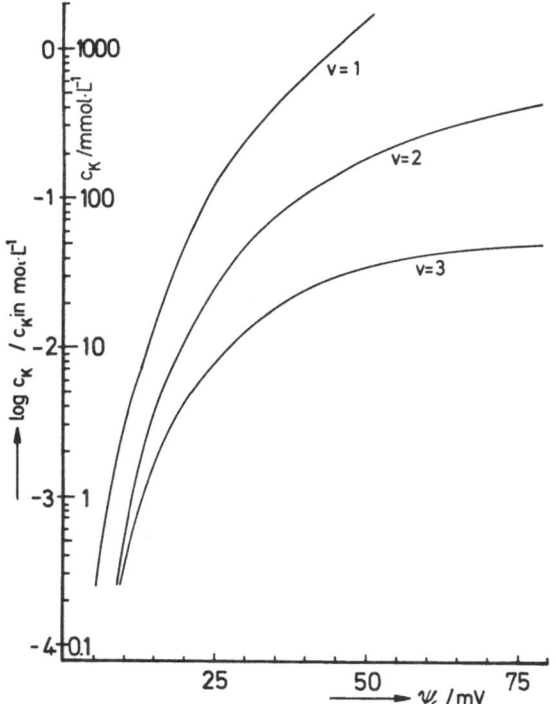

Fig. 4. Critical coagulations concentrations calculated by Equation (1) (Hamaker constant $A = 10^{-20}$ J)

of the gegen ions and the surfactant anions on the water structure shifts the maximal c_K from CsCl to LiCl as the SDS/styrene ratio increases.

Anions affect the water structure even more strongly than cations [28, 29]. The coagulation concentrations of electrostatically stabilized dispersions increase with increasing structure temperature of the anions ($F^- \rightarrow ClO_4^-$, Table 3). The increasing water structure breaking of the anions promotes the hydration of the sodium ions and decreases their concentration in the Stern layer. The increased ψ_δ then requires a higher gegen ion concentration to coagulate the system.

Conclusion

The stability of latex dispersions toward salts is governed by two kinds of structural effects: the structure of the surfaces and the structure of the solvent. Latices with smooth surface and pure electrostatic stabilization and latices with a corona of hairs (macromolecules radiating away from the surface as loops or tails) may be considered as the extreme cases. Intermediate states are latex particles with strongly adsorbed surfactant anions and "hardcore" particles surrounded by an "gel-like" envelope of macromolecules. The different coagulation power of the cations and anions is related to their influence onto the water structure.

Acknowledgement

We thank Prof. Dr. E. O. Schulz-DuBois and his group (Institut für angewandte Physik, Kiel University) for dynamic light scattering and AWPS measurements. We are grateful to the "Fonds der chemischen Industrie" for financial support.

References

1. Zimehl R, Lagaly G (1986) Colloid Surfaces, in preparation
2. Goossens JWS, Zembrod A (1979) Colloid Polym Sci 257:437
3. Sasaki S (1984) Colloid & Polymer Sci 262:406
4. Bensley CN, Hunter RJ (1983) J Colloid Interf Sci 92:48
5. Harding IH (1985) Colloid & Polymer Sci 263:58
6. Yates DW, Goodwin JW, Ottewill RH (1977) J Colloid Interf Sci 62:356
7. Harding IH, Healy TW (1982) J Colloid Interf Sci 89:185
8. Bensley CN, Hunter RJ (1982) J Colloid Interf Sci 88:546
9. Goodwin IW, Hearn I, Ho CC, Ottewill RH (1974) J Colloid Polym Sci 252:464
10. Schulz-DuBois EO (ed) (1983) Photon Correlation Techniques in Fluid Mechanics, Springer, Berlin
11. Schätzel K, März J (1984) J Chem Phys 81:2482
12. van den Esker MWJ, Pieper JHA (1975) (eds) van Olphen H, Mysels KJ, Enriching Topics from Colloid and Surface Science, Thorex, La Jolla, p 175
13. Bijsterbosch BH (1978) Colloid & Polymer Sci 256:343
14. Ito K, Ise N, Okubo T (1985) J Chem Phys 82:5732
15. Vanderhoff JW (1980) Pure Appl Chem 52:1263
16. de Rooy N, de Bruyn PL, Overbeek JThG (1980) J Colloid Interf Sci 75:542
17. Kitahara A, Ushiyama H (1973) J Colloid Interf Sci 43:73
18. Duckworth RM, Lips A (1978) J Colloid Interf Sci 64:311
19. Overbeek JThG (1982) Adv Colloid Interf Sci 16:17
20. Tagawa M, Inoue T, Ogi N, Nakagaki M (1985) Colloid & Polymer Sci 263:406
21. Bibeau AA, Matijević E (1973) J Colloid Interf Sci 43:330
22. Stone-Masui J, Watillon A (1975) J Colloid Interf Sci 52:479
23. Bagchi P, Gray BV, Birnbaum SM (1979) J Colloid Interf Sci 69:502
24. Overbeek JThG (1980) Pure Appl Chem 52:1151
25. Napper DH (1983) Polymer Stabilization of Colloidal Dispersions, Academic Press, London
26. Lyklema J, de Wit JN (1975) J Electroanal Chem 65:443
27. Bernal JD, Fowler RH (1933) J Chem Phys 1:515
28. Luck WAP (1978) Progress Colloid Polym Sci 65:6
29. Luck WAP (1984) (ed) Belford G, Synthetic Membrane Processes, Acad Press, New York p 21

Progress in Colloid & Polymer Science, Vol. 72 (1986)

30. Healy ThW, Homola A, James RO, Hunter RJ (1978) Farad
    Disc 65:156
31. Ruckenstein E (1984) J Colloid Interf Sci 99:270

Received December 16, 1985;
accepted June 10, 1986

Authors' address:

Prof. Dr. Gerhard Lagaly
Institut für anorganische Chemie
der Universität Kiel
Olshausenstraße 40
D-2300 Kiel, F.R.G.

Progress in Colloid & Polymer Science　　　Progr Colloid & Polymer Sci 72:37–42 (1986)

Colloidal systems for controlled drug delivery – structure activity relationships*)

H. E. Boddé, T. De Vringer, and H. E. Junginger

Center for Bio-Pharmaceutical Sciences, Division of Pharmaceutical Technology, Gorleaus Laboratories, Leyden, The Netherlands

Abstract: This study deals with crystalline nonionic surfactant-water mixtures having lamellar gel structures. The release of an extremely hydrophilic drug (nicotinamide) from mixtures with various water contents was measured and the results compared with Differential Scanning Calorimetry data. It is concluded that the lamellar gel structure fits a simple pore model in which the interlamellar channels behave as pores, the porosity being determined by the *free* (i. e. *unbound*) water content, the tortuosity by the orientation of the lamellae.

Key words: Lamellar gel structure, drug release, porosity, tortuosity.

Introduction

Colloidal systems have always received keen interest within the pharmaceutical field, and are being developed and used for various purposes, e. g. as dermatological vehicles and drug delivery systems.

A particularly interesting kind of colloidal system, both from a practical and a fundamental point of view, are the hydrophilic creams for intradermal and transdermal drug delivery. They are generally oil-in-water mixtures stabilized by gel structures formed by surfactants. Being among the oldest dosage forms presently used, primarily for topical application, they are nowadays recognized as competing with recently developed dosage forms, such as transdermal delivery systems, based on self-adhesive polymer patches, which tend to draw widespread attention [1, 2].

Hydrophilic creams have a number of attractive qualities which to date have not been fully explored:

– Hydrophilic creams, especially those containing nonionic surfactants, are mostly 'skin-friendly' and may have an advantage over the usually hydrophobic transdermal patches, which tend to irritate the skin due to an occlusive effect when applied over an extended period [3].

– Surfactants contained in the creams may enhance the transdermal penetration of drugs when allowed to diffuse into the skin to a sufficient degree.

– Depending on the oil/water/surfactant ratios hydrophilic creams adopt various gel structures, which are responsible for technologically important properties such as viscoelasticity, swellability, solubilizing capacity and restricted permeability.

The above mentioned properties are of considerable practical interest, e. g. in pharmaceutical or cosmetical applications.

For example, it seems likely that the gel structures within hydrophilic creams can be used in a systematic way for the controlled release of drugs via the transdermal route, or for the regulation of the water balance inside the skin. In this paper, the emphasis lies on mixtures of nonionic surfactants, crystalline cosurfactants and water, all basic ingredients of a number of oil-in-water creams. These surfactant/water mixtures may be considered as simplified versions of the oil-in-water creams from which they are derived and with which they share many properties. Over a wide range of water concentrations they adopt lamellar, crystalline gel structures. In the lamellar structure the surfactants and

---

*) Lecture presented during the 32nd Annual Meeting of the Kolloid-Gesellschaft, Berlin October 2–4, 1985.

cosurfactants form platelike lamellar aggregates stacked in a repetitive pattern, whereby lipophilic bilayers alternate with hydrophilic layers which contain interlamellarly inserted water [4].

Generally, a drug permeating through a lamellar gel network may follow an interlamellar or translamellar route, depending on local rates of diffusion and partition. Extremely liphophilic drugs will probably be trapped inside the lipophilic bilayers [5]; extremely hydrophilic drugs most likely permeate through the hydrophilic regions between the lamellae, while amphilic drugs may move both between and across the lamellae. In the latter case, various interesting release patterns are theoretically expected, depending for example on drug lipophilicity [6]. In order to clarify the role of the hydrophilic regions within the lamellar structure with respect to drug permeation, we studied the release of nicotinamide, an extremely hydrophilic nonionic drug lacking acid-base behaviour, from crystalline systems with lamellar structures.

In order to find out how the water structure inside the systems influences drug release, the release data were compared with data obtained by Differential Scanning Calorimetry [7].

Experimental

Nonionic drug-containing lamellar systems were prepared by adding various amounts of a 5% aqueous solution of nicotinamide (pyridine −3− carbonamide) to a 1:1:1 mixture (weight ratio) of PGM_{20} (polyoxyethylene(20)glycerylmonostearate; Tagat S_2, Goldschmidt), cetyl- and stearyl alcohol (Lorol C_{16} and C_{18}, Henkel) mixing at 70 °C, subsequently cooling off to 25 °C and finally allowing the mixtures to stand for one month. For the sake of simplicity, these mixtures will henceforth be referred to as "creams". The total water contents of the creams ranged from 15.6 up to 42.4 vol %. The volume percentages were calculated from the weight percentages, using a method described elsewhere [8]. The lamellar structures of the creams were confirmed by Small Angle X-ray Diffraction (SAXD) [8].

Cumulative release of nicotinamide was measured in vitro (using UV absorbance at 261 nm) from a 5 mm thick cream layer via a cellulose membrane with a surface area of 3.14 cm² to an aqueous sink at 25 °C in a simple two compartment set-up. All experiments were done in triplicate. In order to avoid complications due to diffusion of buffer components and possible disturbance of the gel structures within the creams, no buffers were used. Identically prepared creams, only without the drug, were analyzed by Differential Scanning Calorimetry using a Mettler DSC30 calorimeter (TA 3000). The samples were cooled down from 25 °C to −60 °C at a cooling rate of 5° min⁻¹.

Results and discussion

The results of the release experiments are compiled in Figure 1, which contains plots of the cumulative nicotinamide release versus the square root of time for all creams studied (the 19.3 vol % plot is omitted for the sake of clarity); the straight lines are obtained with linear regression. Standard deviations in release data are about 5 %.

In all cases there is complete linearity, except for the 15.6 and 21.1 vol. % plots, which slightly deviate from linearity, probably due to water uptake by the creams during the release experiments. The linearity of the release plots agrees with general formulae [9, 10] for short term release from a semi-infinite source to a perfect sink, based on Fick's diffusion laws.

In order to explain the release data in more detail, the structural properties of the creams are now taken into account. From extensive studies [4, 7, 8], using SAXD and DSC, it is known that at water concentrations ranging from 0 up to at least 55 % these creams adopt lamellar structures, with the cetyl and stearyl chains constituting the lipid bilayers, and the strongly hydrated oxyethylene moieties sticking into the intermediate water layers [Fig. 2]. A close relationship has been found between the interlamellar distance and the water content increasing from about 20 % to about 55 % w/w [8]. Structural data [8] furthermore strongly suggest that the lamellar gel structure is built up of randomly oriented small crystalline domains, each consisting of a number of lamellae, stacked upon each other.

Now we can derive a simple *pore model* based upon the following assumptions:

i) The hydrophilic drug nicotinamide follows an *interlamellar* diffusion pathway with the interlammelar oxyethylene-water layers acting as 'channels' or 'pores' (Fig. 3). We therefore define an *intrinsic diffusion coefficient* D_o for diffusion inside these 'pores'.

ii) Due to the presence of strongly hydrophilic oxyethylene moieties, it is most likely that the water inside the 'pores' only partly contributes to the drug permeation. Hence a *porosity* ε, based on the water content, is defined as follows:

$$\varepsilon = w - w^* \tag{1}$$

where:

w = total water content
w^* = 'impermeable' water fraction

iii) The drug follows a tortuous path, determined by local variations in the orientation of the lamellae

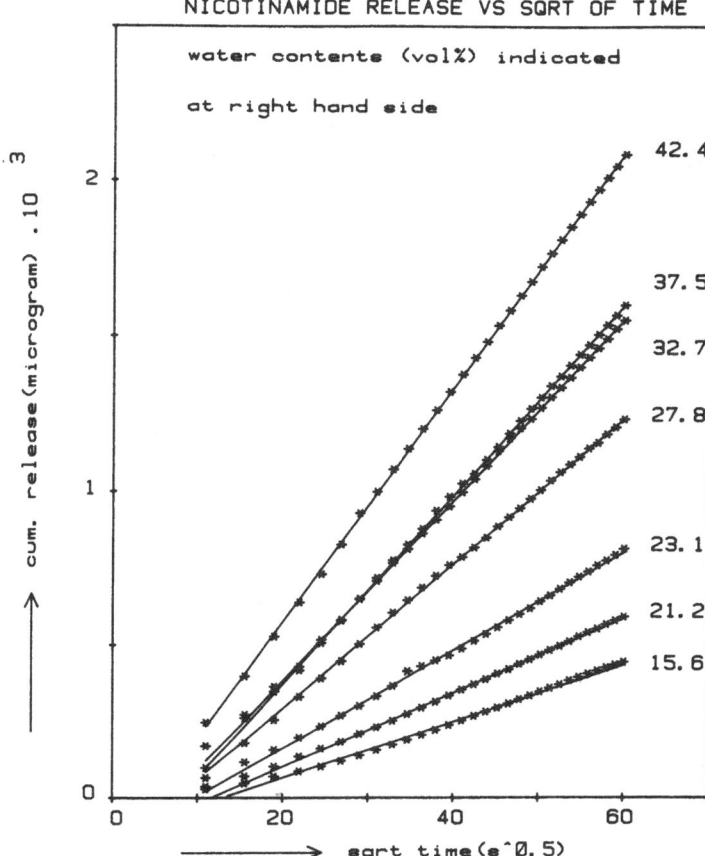

Fig. 1. Root of time plots of the cumulative nicotinamide release from creams with various water contents

(Fig. 3). Let the orientation of a (stack of) lamella(e), with respect to the main direction of transport be characterized by an angle θ; the *tortuosity* may then be defined as the mean square cosine $\langle \cos^2 \theta \rangle$. Assuming *random orientation*, we have:

$$\langle \cos^2 \theta \rangle = 0.5.$$

iv) We now express the *effective diffusion coefficient* D for drug diffusion within the creams as:

$$D = D_o \cdot \varepsilon \cdot \langle \cos^2 \theta \rangle \qquad (2)$$

and write the short term cumulative release as [9, 10]:

$$Q = 2 C_o A \left[\frac{Dt}{\pi} \right]^{1/2} \qquad (3)$$

where:

Q = cumulative amount of drug released
C_o = initial drug concentration in the source
A = interfacial area.

With these assumptions, we obtain a linear relationship between the *squared slope* of a release plot and the corresponding *water content* of the cream:

$$\left[\frac{dQ}{dt^{1/2}} \right]^2 = KD_o \, (w - w^*) \qquad (4)$$

where

$$K = \frac{4 \, C_o^2 A^2 \, \langle \cos^2 \theta \rangle}{\pi}.$$

To put this model to the test, the squared slopes of the release profiles were plotted against the water contents (Fig. 4) and fitted with the model (Eq. 4).

Curve fitting yielded:

$$D_o = (3.6 \pm 0.3) \cdot 10^{-6} \text{ cm}^2 \text{ s}^{-1}$$

$$w^* = 19 \pm 2 \text{ vol \%}.$$

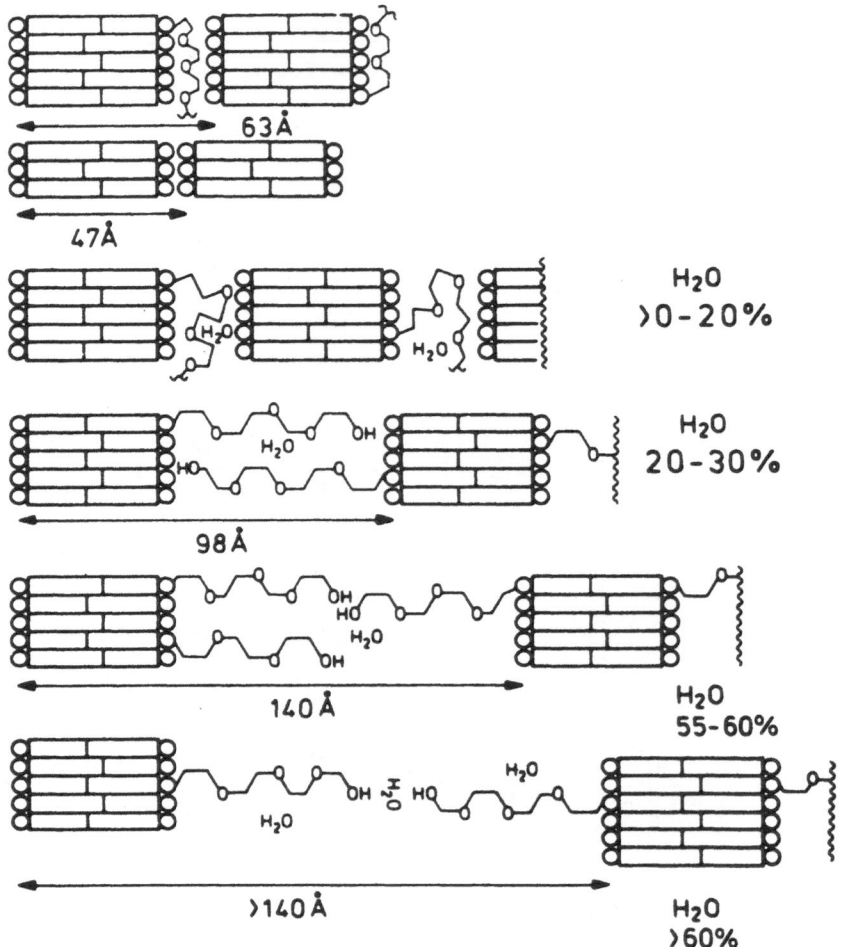

Fig. 2. Lamellar gel structures within creams, visualizing the dependence of the interlamellar distance on the water content

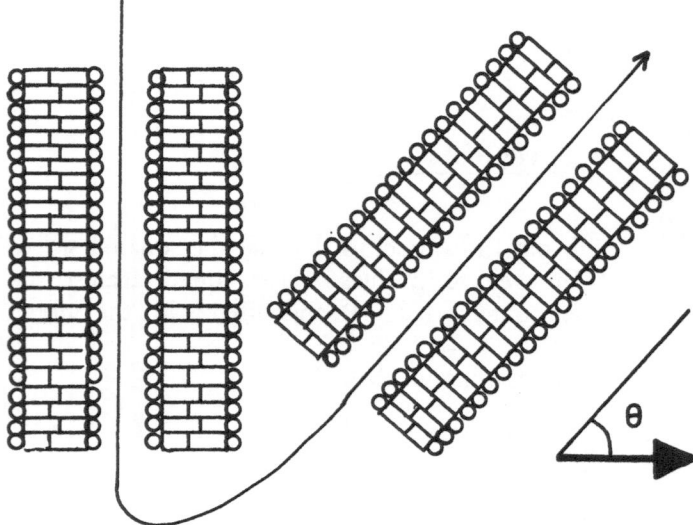

Fig. 3. Schematized tortuous diffusion pathway through variably oriented interlamellar 'channels'. The oxyethylene moieties are omitted for clarity

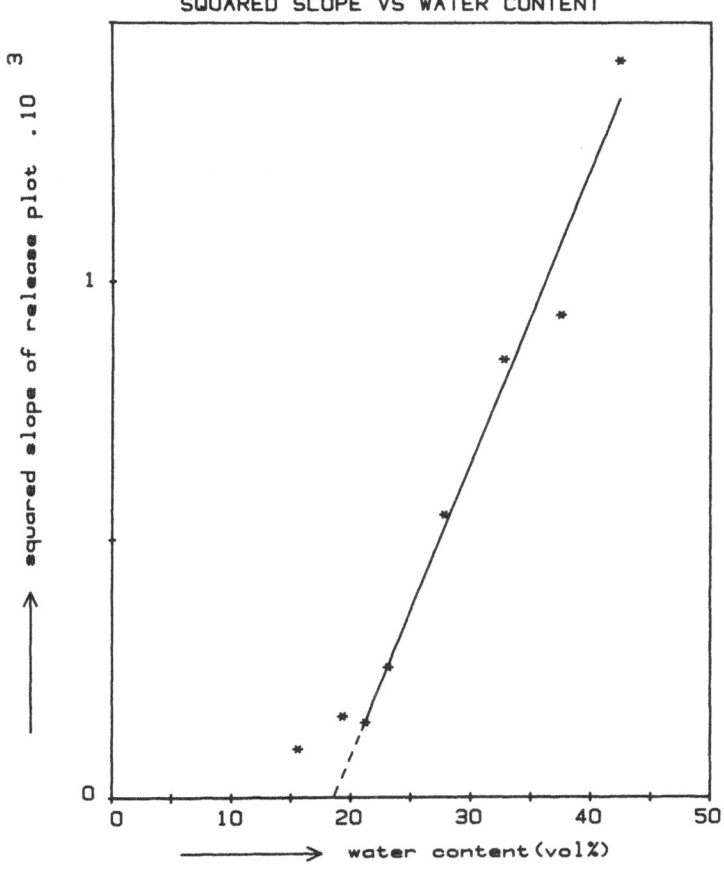

Fig. 4. Squared slopes of the nicotinamide release plots vs. the water contents of the creams

The value for D_o seems quite reasonable for the diffusion of a low molecular compound through an aqueous gel phase.

From the results obtained with Differential Scanning Calorimetry, the freezing enthalpies were calculated [7]; these are plotted versus the water contents in Figure 5. Linear extrapolation towards the composition axis of this plot yields a value of 19 % w/w (about 18 vol %) for the 'non-freezing' water fraction within these systems [7], which value compares rather well with the value found for the 'impermeable' water fraction, w^* (is 19 vol %).

Since the 'non-freezing' water fraction is thought to constitute a structured hydration layer, firmly bound to the oxyethylene chains within the hydrophilic gel phase, it is indeed quite likely that it does not contribute to the drug permeation. Note that for water contents below 19 % the proposed model degenerates and incorrectly predicts immobilization of the drug, because the "porosity" parameter ε vanishes. At these low water contents drug diffusion most likely follows a

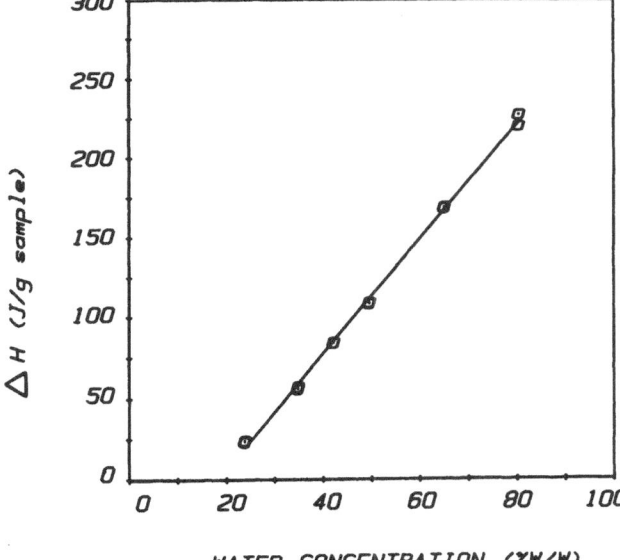

Fig. 5. Freezing enthalpies plotted against the water contents of the creams

different mechanism and probably takes place along the partly hydrated oxyethylene chains.

We finally conclude that with respect to a hydrophilic drug the lamellar gel structure in *o/w* creams behaves as a 'porous' system, the 'porosity' being related to its 'free' water content, the 'tortuosity' to the orientation of the lamellae.

References

1. Chandrasekharan SK, Bayne W, Shaw JE (1978) J Pharm Sci 67(10):1370–1374
2. Curry SH, Kwon HR, Perrin JH, Culp JR, Pepine CJ, Yu W (1984) The Lancet 1297
3. Hurkmans JFGM, Boddé HE, Van Driel LMJ, Van Doorne H, Junginger HE (1984) Br J Dermatol 112:461–462
4. Junginger HE (1984) Pharm Weekbl Sci Ed 6:141–149
5. Wahlgren S, Lindstrom AL, Friberg SE (1984) J Pharm Sci 73(10):1484–1486
6. Boddé HE, Joosten JGH (1985) Int J Pharm 26:57–76
7. De Vringer T, Joosten JGH, Junginger HE, Coll & Polym Sci, submitted
8. De Vringer T, Joosten JGH, Junginger HE (1984) Coll & Polym Sci 262:56–60
9. Crank J (1956) The Mathematics of Diffusion, Academic Press, London New York
10. Higuchi T (1963) J Pharm Sci 52:1145

Received November 28, 1985;
accepted May 20, 1986

Authors' address:

Dr. H. E. Boddé
Center for Bio-Pharmaceutical Sciences
Division of Pharmaceutical Technology
Gorlaeus Laboratories, P. O. Box 9502,
2300 RA Leyden, The Netherlands

Progress in Colloid & Polymer Science

Progr Colloid & Polymer Sci 72:43–50 (1986)

A new method for detecting interactions between macroscopic bodies: Forces between fused silica plates bearing adsorbed polystyrene layers in cyclohexane*)

P. Belouschek and S. Maier

Institute of Physical and Theoretical Chemistry of the Universtiy of Essen, F.R.G.

Abstract: A new method is presented which is suitable for the determination of both attractive and repulsive interaction forces between plates bearing adsorbed polymer layers immersed in a liquid medium. The interaction forces between polystyrene layers adsorbed onto fused silica plates in cyclohexane have been measured at $T = 293$ K. The force-distance diagrams obtained for pure cyclohexane and the adsorbed polystyrene layers are characterized by an extensive range of van der Waals attraction with a minimum followed by an increase of repulsion when coming to lower film thicknesses.

Key words: Interaction forces, macroscopic system, polystyrene layers, cyclohexane, van der Waals attraction.

1. Introduction

The stability of dispersed systems is controlled by a combination of interparticle forces. Therefore, many attempts have been made to investigate these interaction forces between macroscopic interfaces of well-defined geometry, e. g. flat plates or spheres, and to transfer the results to interacting colloidal particles. Thin liquid films are also very useful model systems in this respect [1]. A review of techniques developed for direct measurements of interaction forces between solids as a function of the distance of separation of their surfaces has only recently been reported by K. B. Lodge [2].

While many investigations have been carried out to study the electrostatic and van der Waals interaction forces, results concerning direct investigations of structural and steric forces are meagre. In particular, experimental studies of forces between surfaces bearing adsorbed polymer layers are scarce, in spite of their theoretical and practical interest.

The majority of the published papers deals with repulsive effects of sterically-stabilized free polymer films [3, 4], and with direct detection of repulsive forces between two solids immersed in a polymer solution [5–7]. To our knowledge, experimental investigations of both attractive and repulsive interaction forces between polymer layers have only been carried out by J. Klein and coworkers, who measured the interaction forces between smooth mica surfaces bearing adsorbed macromolecules in liquid media [8, 9].

In this paper we present a new method for determining experimentally both the attractive and the repulsive interaction forces between two solid bodies. The functioning of the apparatus presented below has already been checked by examining the attractive and repulsive interaction forces between two fused silica plates with interjacent layers of aqueous electrolyte solutions [10].

The aim of this paper is to show that the new method is suitable for measuring attractive and repulsive interaction forces between two polymer layers adsorbed on highly-polished plates immersed in a liquid medium. This will be exemplarily presented by investigating the interaction forces between two fused silica plates bearing adsorbed layers of polystyrene in

---

*) Lecture presented during the 32nd Annual Meeting of the Kolloid-Gesellschaft, Berlin October 2–4, 1985.

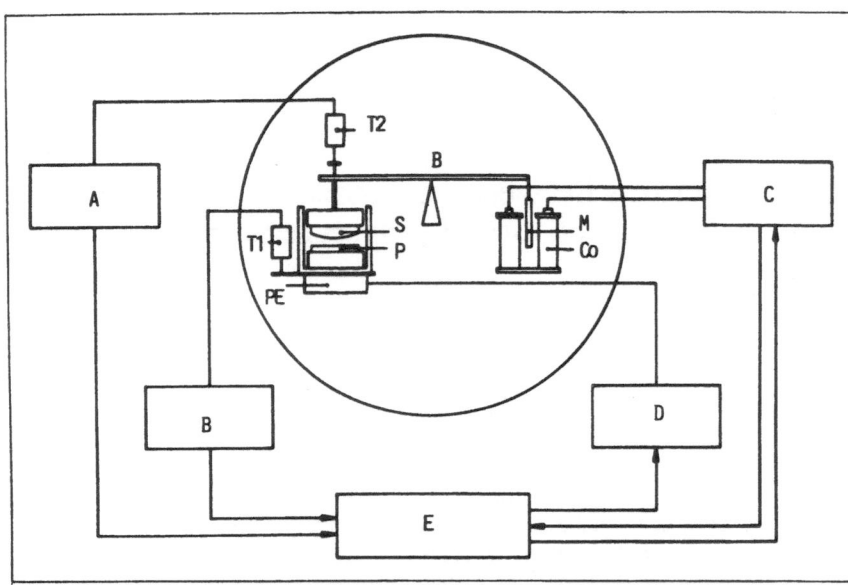

Fig. 1. Scheme of the experimental device: *A*: measuring bridge 2; *B*: measuring bridge 1; *C*: current control; *D*: voltage control; *E*: microcomputer. *T*1 and *T*2 displacement transducer *PE* piezoelement: *S*: spherically formed plate; *P*: planar plate; *B*: balance system; *M*: magnetic rod; *Co*: electromagnetic coil

cyclohexane. Cyclohexane is a poor or worse-than-θ solvent for polystyrene, so that in this experiment attraction might be expected.

2. Experimental

2.1 Method

An experimental device is presented which enables the investigation of both attractive and repulsive interaction forces acting in a planar/spherical plate system. The method is described in more detail elsewhere [10]. The apparatus presented in the following is based on an experimental device which was developed the determination of only repulsive interactions [11, 12].

The schematic presentation of the experimental device is shown in Figure 1. The planar plate (*P*) is fixed to the bottom of a container which can be filled with the liquid system to be studied. The container is fitted to a piezoelement (*PE*). By varying the voltage of the piezoelement the container — and consequently the planar plate too — are moved in a vertical direction. The distances covered are recorded by the displacement transducer (*T*1).

The upper, spherically formed plate (*S*), being centered opposite to the planar one, is attached to one end of a scale-beam. At the other end of the scale-beam a magnetic rod (*M*) is mounted dipping into an electromagnetic coil (*Co*). The balance system can be deflected by varying the current of the electromagnetic coil. The distances thus covered by the spherically formed plate are registered by the second displacement transducer (*T*2).

The adjustment of the electromagnetic coil and of the voltage of the piezoelement as well as the registration and the processing of measuring signals is controlled by a microcomputer. The whole set-up is placed in a thermostated vessel.

For determining the interaction forces as a function of defined separations, a plate separation is adjusted by deflection of the scale-beam where interaction forces can be excluded. Thereon the planar plate is made to approach the spherically shaped plate by means of the piezoelement, in order to adjust a defined film thickness. If interaction between the surfaces arises, the position of the upper, spherically formed plate is changed appropriately. The change of position of the beam of balance which may be positive or repulsive is a measure of forces acting between the two plates by a compensating procedure. The differences in current are converted into the applied force (*K*) by means of a calibration factor. It should be pointed out that the interaction force-related to a defined film thickness is only derived from stable states of the balance system, after position changes of the scale-beam due to attraction or repulsion between the plates has been taking place.

By repeating this procedure iteratively for the distance range to be carried out the corresponding force-distance diagram of interactions between the two plates is obtained.

2.2 Materials

In this work we used plates made of fused silica glass (Herasil I of Heraeus Quarzschmelze GmbH, F.R.G.). The spherically formed plate has a curvature radius (*R*) of 1 m. The plate surfaces are highly-grade polished and display a residual micro-roughness of about 1–2 nm [13, 14]. In order to obtain a reproducible state of hydrophilicity of the fused silica surfaces, the plate were placed in boiling bi-distilled water for about four weeks, subsequently wetted with hot chromic acid, and then heated up to a temperature of about 400 K.

The polymer used was polystyrene with a molecular weight of $M_W = 6 \cdot 10^5$ and a molecular weight distribution of $M_W/M_n \leq 1.10$ (Pressure Chemical). To cover the silica surfaces with polystyrene layers they were placed in a solution that was prepared of 10 mg polystyrene in 1 l cyclohexane (Uvasol of Merck, F.R.G.). After about 15 hours the polymer solution was sucked off and subsequently exchanged by pure cyclohexane. The latter procedure was repeated a number of times in order to ensure that all the soluble polymer is removed from the liquid phase in the container.

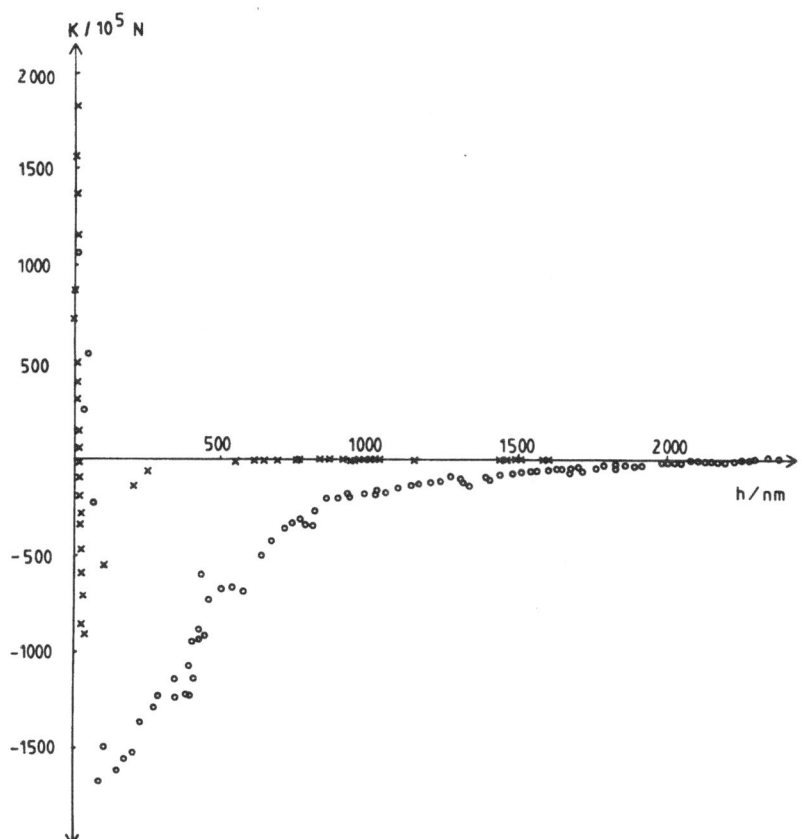

Fig. 2. Force-distance diagram: ×: cyclohexane; O: polystyrene; $T = 293$ K

3. Results and discussion

3.1 General considerations

The interaction forces in pure cyclohexane and in cyclohexane with the adsorbed polystyrene layers are presented in Figure 2. It is evident that both force-distance diagrams are characterized by an extensive range of attraction, with a minimum followed by a drastic increase of repulsion when approaching shorter distances. The minimum of interaction obtained for the polystyrene layers is both more pronounced and also shifted to larger thicknesses than that found for pure cyclohexane.

It is common practice to discuss the interactions between colloidal particles in terms of the scalar (free) energy. This has the advantage that the total energy of interaction (V) can be expressed by the simple addition of the individual contributions. When determining the interactions depending on distance in a planar/spherical system, the energy of interaction can generally be evaluated from the experimental force according to the Derjaguin approximation $V = K/2\pi R$ [15].

3.2 Cyclohexane

As may be inferred from Figure 2, the minimum of the interaction forces occurs at a distance of about 20 nm, followed by a range of repulsive interactions for lower film thickness to be attributed to structural (solvation) forces. The latter are due to the overlapping of multimolecular solvation layers adjacent to the surfaces. Thus, for any plate separation the total interaction energy $(V_{t, h})$ may be expressed by

$$V_{t, h} = V_{vdw} + V_s \qquad (1)$$

with V_{vdw} denoting the van der Waals interactions, and V_s the component due to structural (solvation) interactions.

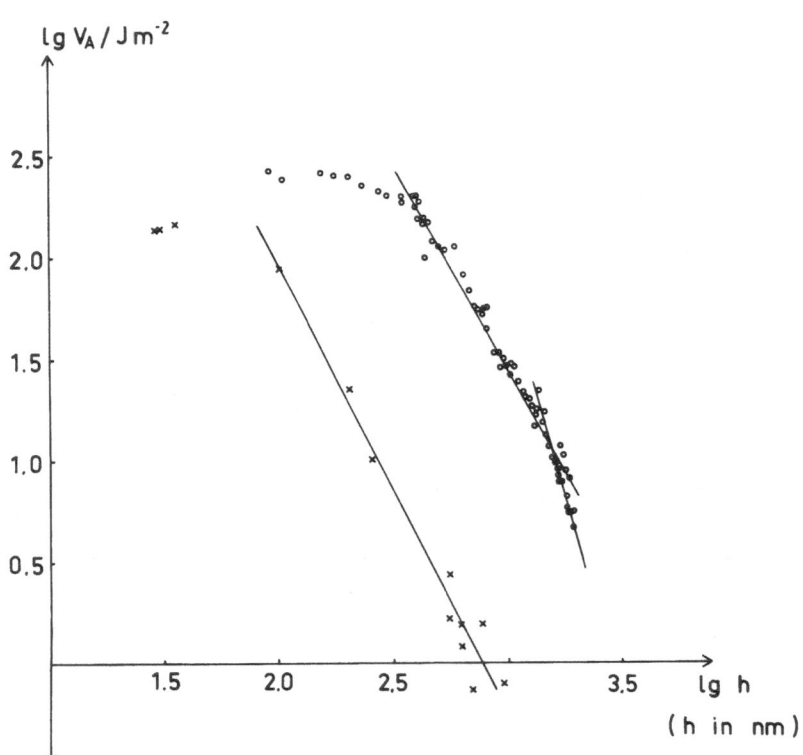

Fig. 3. Logarithm of the interaction energy (V_A) against the logarithm of the distance (h): ×: cyclohexane; O: polystyrene; $T = 293$ K

On the basis of microscopic theories [16, 17] the van der Waals interaction energy per unit area for two semi-infinite parallel plates at distance h is given by

$$V_{vdw} = - A/12\pi h^2 \qquad (2a)$$

if for not too large plate separations the electromagnetic retardation effects are neglected. In the case of electromagnetic retardation of van der Waals attraction it is

$$V_{vdw} = - B/h^3 . \qquad (2b)$$

A and B are the Hamaker constants of the system.

By plotting the logarithm of the experimental V in the range where V_s is negligible against the logarithm of the corresponding plate separation h, a fairly straight line is obtained which enables the determination of the Hamaker constant provided the experimental distance dependence complies with the theoretical prediction (see Fig. 3). For pure cyclohexane the distance dependence ($1/h^2$) is in good agreement with the theory of non-retarded van der Waals interaction. Based on the evaluation of V according to the Derjaguin approximation the Hamaker constant of the

regarded system (quartz (q)/cyclohexane (cy)/quartz (q)) is determined to $A_{(q, cy, q)} = 1.3 \cdot 10^{-15}$ J. The Hamaker constant of the present system may be evaluated theoretically according to the simple expression

$$A_{(q/cy/q)} = (\sqrt{A_q} - \sqrt{A_{cy}})^2 \qquad (3)$$

with the individual Hamaker constants A_q and A_{cy} being tabulated [18]. The Hamaker constant calculated from Equation (3) ($A_{th} = 19.7 \cdot 10^{-19}$ J) is much smaller than that obtained from experiment. This difference indicates that for the present planar/spherical plate system the interactions cannot simply be transferred to plane parallel interactions by means of the Derjaguin approximation. It may rather be assumed that the forces in the present planar/spherical plate system act as if they were obtained by a plane parallel plate system. In this case the Hamaker constant is corrected to $A_{cor} = 7.4 \cdot 10^{-19}$ J which is in good agreement with the theoretical one.

With regard to Figure 3, attention should be called to the fact that the straight line obtained for pure cyclohexane does not extend to the minimum of the force-distance diagram. The deviations from the straight line clearly indicate that already before getting to the minimum, the attractive van der Waals interaction is superposed by an additional interaction, which may be attributed to overwhelming multimolecular solvation layers of cyclohexane adjacent to the fused silica plates. Thus, the repulsion which arises at shorter plate distances cannot be an artefact by the micro-roughness of the polished surfaces. In that case the straight line in the diagram of Figure 3 should extend exactly to the minimum value of interaction since a repulsive effect aroused by surface roughness would occur abruptly at the minimum of the K:h digram.

Based on theoretical and experimental data [19] the distance-dependence of the structural (solvation) term is chosen to be of an exponential form in a first approximation. Thus, the structural force acting in a planar/spherical plate system depends on distance as

$$K_h = 2\pi R / n_s \, \Pi_{s,o} \qquad (4)$$

with the structural disjoining pressure

$$\Pi_{s,h} = \Pi_{s,o} \, e^{-n_s h}. \qquad (5)$$

$\Pi_{s,o}$ is the structural disjoining pressure at $h = 0$, and $1/n_s$ the decay length. When plotting $\ln K$ against h, a fairly straight line is obtained which allows the evaluation of $\Pi_{s,o}$ and n_s. For cyclohexane it renders $\Pi_{o,s} = 1.2 \cdot 10^{-6} \, \mathrm{Nm}^{-2}$ and $1/n_s = 13$ nm. The free excess energy of structural interaction is given by

$$V_s = \Delta F_s^E = - \int_\infty^h \Pi_{s,h} \, dh'. \qquad (6)$$

The free excess energy needed to expel the liquid layers from the space between the surfaces by approaching them down to a distance of $h = 2\delta'$, where the first adsorbed layers on both the surfaces will be left, is determined at $\Delta F_{2\delta'}^E = 1.8 \cdot 10^{-3} \, \mathrm{J/m^2}$ contributing to only a small extent to the free interfacial energy of some 0.2 J/m² [20].

The structural (solvation) interaction of cyclohexane is characterized by a weak but long-range effect, whose physical relevance is constituted by a change of the state of intermolecular vibration of the liquid molecules perpendicular to the surface, namely a reduction of molecular mobility adjacent to solid walls. The latter interpretation is based on investigations of thermal conductance of alkanes and alcohols in thin layers between two fused silica plates [21, 22].

3.3 Polystyrene

Before proceeding to an analysis of the force-distance diagram for polystyrene (*ps*) shown in Figure 2, it is necessary to make some remarks concerning the adsorbance of polystyrene on the fused silica plates which lacks direct information. Based on adsorption studies of polystyrene ($M_w = 6 \cdot 10^5$) onto various silica substrates [23, 24] the adsorbance of (*ps*) may be assumed to be some 5 mg/m² which can be compared with the equilibrium adsorbance of some 3–8 mg/m² onto various types of solid surfaces from cyclohexane at the θ-temperature, $T = 307.5$ K [25].

By carrying out ellipsometrical studies of polystyrene adsorbed onto various surfaces from cyclohexane near the θ-temperature the adsorbed layer thickness is determined at $\delta = 40$–50 nm [26]. It may be concluded from Figure 2 that the minimum of interaction between the fused silica plates bearing polystyrene layers corresponds to a distance of about 90 nm followed by an increase of repulsion with decreasing film thickness. This distance is comparable to 2δ, twice the adsorbed layer's thickness. 2δ is about equal to $4 R_G \cdot R_G$ is the radius of gyration which for polystyrene with $M = 6 \cdot 10^5$ amounts to 21 nm.

Thus, for distances greater than 2δ, it may be assumed that the interaction between the fused silica plates bearing polystyrene layers in the poor solvent cyclohexane is chiefly governed by the van der Waals attraction. Considering again Figure 3 which shows the plot of lg V_A against lg h, it is evident that for polystyrene two straight lines of different slopes are yielded. The distance dependences of van der Waals attraction are determined to $(1/h^{1.9})$ and $(1/h^{2.9})$ indicating the existence of retarded and non-retarded interactions. The van der Waals attraction is found to be retarded at distances larger than 1000 nm, in agreement with results obtained by Wittmann et al. [26].

Based on the result that the present planar/spherical plate system acts plane parallel, the non-retarded Hamaker constant is found to be $A^* = 4.7 \cdot 10^{-17}$ J, and the retarded constant $B = 2.2 \cdot 10^{-26}$ Jm. The Hamaker constant theoretically evaluated from the individual constants [18] on the basis of Equation (3) is much smaller than that experimentally obtained. The theoretical Hamaker constant of the system (*q/ps/cy/ps/q*) is given to $A_{th}^* = 1.9 \cdot 10^{-20}$ J. For understanding this discrepancy it has to be stimulated that A^* is derived from the experimental van der Waals forces adopting one unit of area of the fused silica surface as the reference. A^* is a composed Hamaker constant which contains

Progress in Colloid & Polymer Science, Vol. 72 (1986)

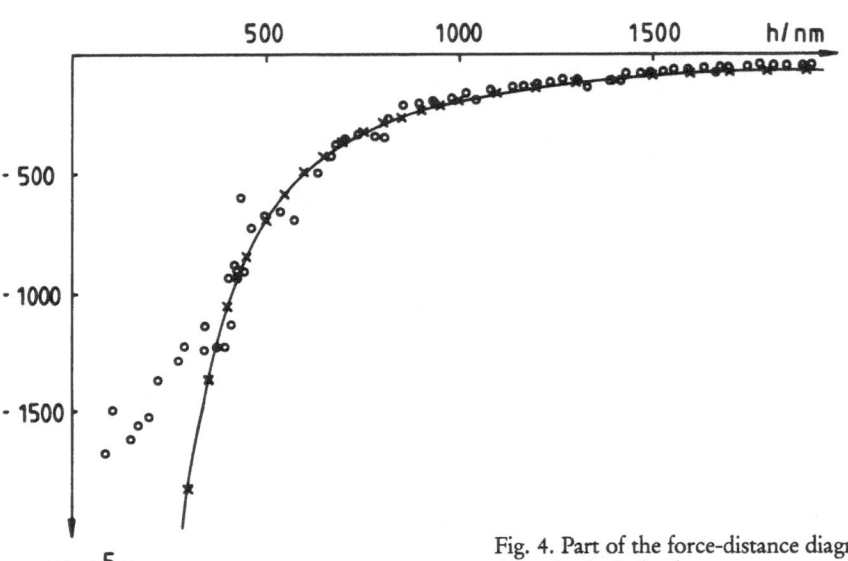

Fig. 4. Part of the force-distance diagram for polystyrene: O: experimental values; —×—: theoretical values

the contribution of the individual materials. When comparing the surface of the fused silica plates with that of the adsorbed polystyrene layers it may well be that due to the protrusion of loops and tails into the solution — producing a kind of artificial surface roughness — the effective polymer layer surface area is much greater than the corresponding area of interaction of the fused silica plates. This means that in reality the number of interacting atomic units of polystyrene is much greater than would be expected on the basis of effective silica area. Since A^* is calculated from the experimental of V, an artificial increase of the Hamaker constant ensues. This view is supported by van Bree et al. [27] who pointed out that surface roughness can easily raise the van der Waals interaction forces to an extensive degree.

As may be inferred from Figure 4, which displays the experimental force-distance diagram of polystyrene in combination with the theoretical curve of nonretarded van der Waals interaction in the range of about 2000 nm to 90 nm, the experimentally obtained interaction force for $h = 300 - 90$ nm is attractive but does not follow the van der Waals theory. It is obvious that the deviations from the experimental graph are always positive, indicating the superposition of repulsive interactions which, for example, may be due to polystyrene tails protruding into the solvent. It is interesting to note that according to the theory of steric interaction, the repulsion starts at plate separations of about $h = 300$ nm (see Fig. 5). Thus, in the present

case, bridging effects may be neglected in a first approximation.

It is well known that the steric repulsion of overlapping adsorbed polymer layers is governed by the number of loops and/or tails. In order to obtain some information on the number of loops and tails per unit area, required for estimating the area increase of the interacting polystyrene surface, we undertake a semiquantitive interpretation of the repulsive effect assuming tails and/or loops of equal legth. Hesselink et al. [28] derived the following expression

$$V_{vr} = 2 \, vkT \, W_{(i, h)} \qquad (7)$$

for the free energy due to volume restriction of an assembly of average tails or loops. i is the number of segments in a tail or loop, v the number of tails or loop per unit area. $W_{(i, h)}$ is to be found in tables [28]. The osmotic contribution, due to the overlap of polymer layers, is introduced as

$$V_m = 2 \left(\frac{2\pi}{g}\right)^{3/2} (a^2 - 1) \, kT \, v^2 \, \langle r^2 \rangle \, M_{(i, h)} \qquad (8)$$

by the same authors. a is the linear expansion parameter, $\langle r^2 \rangle$ the conformational parameter. $M_{(i, h)}$ is also tabulated [28].

We carried out the measurements near θ-conditions, so that a in Equation (8) may be assumed to be 1 and in a first approximation the osmotic contribution

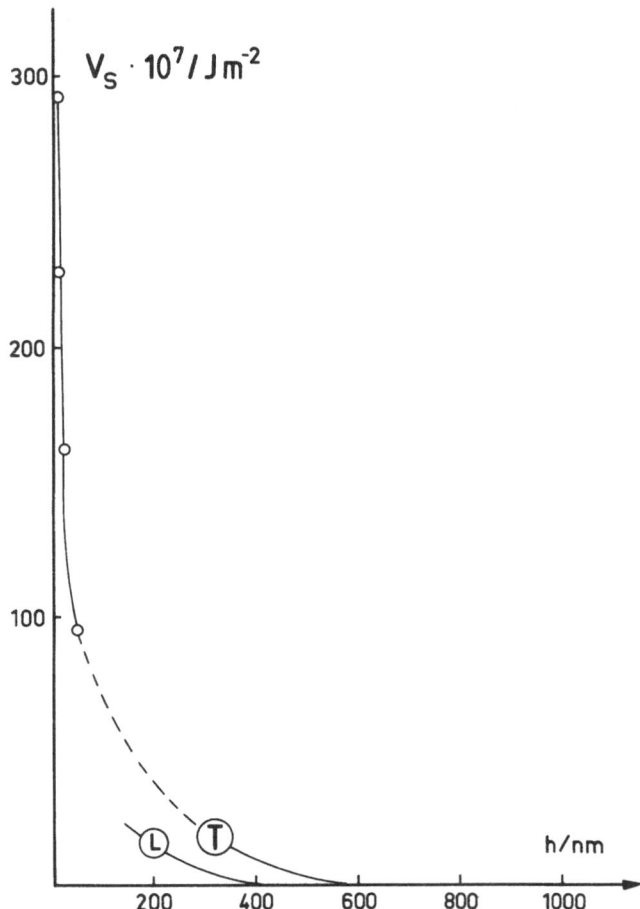

Fig. 5. Comparision of theoretical and experimental repulsive interaction energy between polystyrene layers: O: experimental values; —⊘— theoretical values valid for tails; —⊕— theoretical values valid for loops both calculated according to Equation (9)

is supported by Takahashi et al. [29] which by comparing the root mean-square extension of the adsorbed layer and the average size of loops (P_B) of polystyrene show that above $M_w = 5 \cdot 10^5$ the calculated extension of P_B is nearly a constant equal to 12 nm, being much smaller than the adsorbed layer thickness. The latter view is in agreement with modern theories of polymer adsorption developed by Scheutjens and Fleer [30].

Thus, for estimating qualitatively the remarkable increase of van der Waals forces between fused silica surfaces covered with polystyrene in cyclohexane we focus the discussion on the possible contribution of tails sticking into the solvent. For a tail number of some 10^{12} per cm^2 and the assumption of about 100 segments per tail an increase of the interacting polymer surface on the order of magnitude 10 results. When taking into account the contribution of the loops it is plausible that the Hamaker constant as derived from the experimental data is greater than the theoretical one by a factor of some 10 to 100.

4. Conclusion

The results clearly demonstrate that the presented method is suitable for measuring attractive and repulsive interaction forces between macroscopic bodies bearing polymer layers in a liquid medium. The essential advantage of the apparatus is the possibility to determine the interaction force for any given plate separation. Thus, the method enables not only the investigation of steric repulsion but also the van der Waals attraction on the total distance range.

Acknowledgement

The authors gratefully thank Prof. B. H. Bijsterbosch for the very helpful discussion.

(Eq. (8)) to steric stabilisation may be neglected. Thus, the interaction energy of the overlapping adsorbed polymer layers as a function of distance is given by

$$V_{ps} = V_{vr} + V_{vdw}. \qquad (9)$$

Based on the assumed adsorbance of 5 mg/m^2, the number of loops or tails is evaluated at some 10^{12}/cm^2. The corresponding steric interaction energy as a function of plate separation combined with the van der Waals attraction (see Eq. (9)) is presented in Figure 5. As may be concluded from Figure 5, despite the lack of more detailed information, a reasonable fit between theory and experiment is obtained indicating the predominant influence of tails. This qualitative conclusion

References

1. Lyklema J (1981) Ber Bunsenges Phys Chem 85:826
2. Lodge KB (1983) Adv Colloid Interface Sci 19:27
3. Lyklema J, van Vliet T (1978) Faraday Discussion 65:25
4. Sonntag H, Unterberger B, Zimontkowski S (1979) Colloid Polymer Sci 257:286
5. Cain FW, Ottewill RH, Smitham JB (1978) Faraday Discussion 65:33
6. Israelachvili JN, Tandon RK, White LR (1980) J Colloid Interface Sci 78:430
7. Sonntag H, Ehmke B, Miller R, Knapschinski (1982) Adv Colloid Interface Sci 16:381
8. Klein J (1982) Adv Colloid Interface Sci 16:101
9. Klein J, Luckham P (1984) Nature 308:836

10. Belouschek P, Maier S, contributed to Coll & Polym Sci
11. Adlfinger KH, Peschel G (1970) Z Phys Chem, NF 70:151
12. Peschel G, Belouschek P (1976) Progr Coll & Polym Sci 60:108
13. Lichte H (1978) private communication
14. Fa Zeiss, Oberkochen (1985) private communication
15. Derjaguin BV (1935) Koll Zeitschr 69:155
16. Hamaker CH (1937) Physika 4:1058
17. de Boer JH (1937) Trans Faraday Soc 32:11, 21
18. Vincent B (1973) J Coll Interf Sci 42:270
19. Churaev NV, Derjaguin BV (1985) J Coll Interf Sci 103:542
20. Neumann AW (1974) Adv Coll Interf Sci 4:1
21. Belouschek P, Suppa M (1984) (ed) Proc of 3rd Discussion Meeting on Physical Chemistry of Finely Dispersed Systems, Polish Academy of Science, pp 14–19
22. Belouschek P, Suppa M, Sattler U (1985) Z Phys Chem (NF) 146:65
23. Fursusawa K, Yamashita K, Konno K (1982) J Coll Interf Sci 86:35
24. Kawaguchi M, Maeda K, Kato T, Takahashi A (1984) Macromolecules 17:1666
25. Stromberg RR, Tutas DJ, Passaglia E (1965) J Phys Chem 69:3955
26. Wittmann F, Splittgerber H, Ebert K (1971) Z Phys 245:354
27. van Bree JLMJ, Poulis JA, Verhaar BJ, Schram K (1974) Physika, Utrecht 78:187
28. Hesselink F TH, Vrij A, Overbeek J TH G (1971) J Phys Chem 75:2094
29. Takahashi A, Kawaguchi M, Hirota H, Tadaya K (1980) Macromolecules 13:884
30. Scheutjens JMHM, Fleer GJ (1980) J Phys Chem 84:178

Received December 3, 1985;
accepted May 15, 1986

Authors' address:

Dr. Peter Belouschek
Institut für Physikalische und Theoretische Chemie
der Universität Essen
Universitätsstraße 5–7
D-4300 Essen 1, F.R.G.

Progress in Colloid & Polymer Science   Progr Colloid & Polymer Sci 72:51–59 (1986)

Shear induced phase transitions in dilute aqueous surfactant solutions*)

H. Rehage, I. Wunderlich, and H. Hoffmann

Institut für Physikalische Chemie der Universität Bayreuth, Bayreuth, F.R.G.

Abstract: Aqueous solutions of some cationic surfactants show shear induced phase transitions in the highly dilute concentration regime. Rheological data and results of flow birefringence measurements indicate that the micellar structures in these solutions undergo flow induced sol → gel transitions.

Key words: Phase transition, stress relaxation, flow birefringence, surfactant solution, first normal stress difference.

1. Introduction

Shear induced phase transitions have been observed in dilute aqueous solutions of surfactants [1–6]. Sometimes such transitions can occur slightly above the critical micelle concentration (cmc). Under these conditions the mean distance of separation between the micelles is much larger than the dimensions of the aggregates. At higher concentrations, the solutions of these systems behave like gels and exhibit shear thinning behavior [1]. The size and the shape of the aggregates that are present in the dilute surfactant solutions, have been intensively studied with numerous experimental techniques, such as NMR, electric birefringence, flow birefringence, small angle neutron scattering, light scattering and kinetic measurements [1–6]. The results of these investigations point to the existence of globular particles at concentrations slightly above the cmc, and to rod-like micelles for concentrations above a characteristic threshold value c_t [1–7]. These anisometric particles have the general tendency to grow in length with surfactant concentration [8]. At low shear rates or angular frequencies, a solution of these particles exhibits Newtonian flow. Under these conditions, the surfactant systems behave like any dilute solution of colloidal particles and the viscosity

values can be used to obtain information on the size of the particles [8, 9]. At higher shear rates, however, the rheological properties change dramatically. Above a well defined threshold value of the velocity gradient, a supermolecular structure can be formed from the micellar aggregates. This shear induced structure behaves like a gel and exhibits strong flow birefringence [10]. The flow behavior of the shear induced state is very complicated and has not yet been studied systematically. In this paper we discuss new results which have been obtained from rheological measurements and flow birefringence data. We describe the dynamic processes of structural reorganization and we analyse the rheological properties of the shear induced state. The measurements have been carried out on solutions of the surfactant system Tetradecyltrimethylammoniumsalicylate (TTAS) where the phenomenon of shear induced phases can easily be detected.

2. Experimental

Materials and methods

Tetradecyltrimethylammoniumsalicylate (TTAS) we prepared as previously described by ion exchange procedure from TTACl-solutions or by dissolving TTAS which has been synthesized previously [1–6]. Both methods gave identical results. The solutions were left standing for two days in order to reach equilibrium. The rheological properties of the detergent solutions were measured with the Rheometrics Fluids Rheometer RFR 7800 (cone and plate geometry). The experimental equipment for flow birefringence measurements consists of a concentric cylinder type apparatus, which has been described in detail in [1, 11, 12].

---

*) Lecture presented during the 32nd Annual Meeting of the Kolloid-Gesellschaft, Berlin October 2–4, 1985.

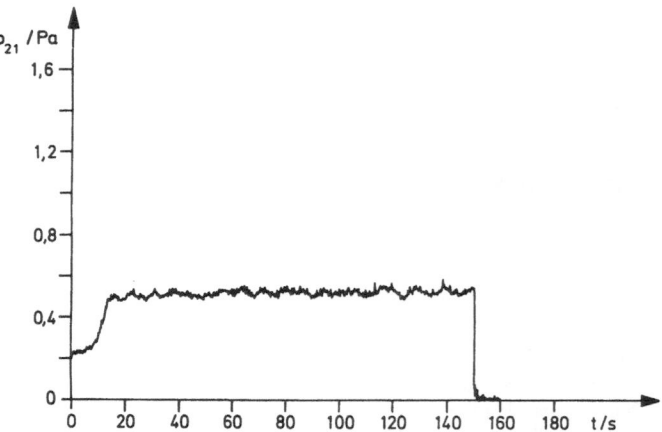

Fig. 1. Dependence of the shear stress p_{21} on time of shearing at $\dot{y} = 200 \, s^{-1}$ (TTAS, $T = 20\,°C$, $c = 2.5 \cdot 10^{-3}$ M)

3. Results

The transient behavior of surfactant solutions can be investigated by measuring time dependent rheological properties. In this type of experiment, a step function shear rate is suddenly applied to the solution. Typical results of these experiments are represented in Figure 1 and Figure 2.

In this experiment, a step function shear rate has been applied at $t = 0$ and the shear stress p_{21} and the first normal stress difference $p_{11} - p_{22}$ have been measured as a function of the shear time. Immediately after the onset of flow we observe a constant shear stress p_{21} indicating that the viscosity follows Newton's law:

$$p_{21} = \eta \cdot \dot{y} \tag{1}$$

The viscosity value of the solution is very low and only slightly higher than the viscosity of the pure solvent. After a short shearing time (≈ 10 s) the stress increases and finally a steady-state value is reached. In the regime of increasing shear a flow induced phase transition takes place and a new structure with an enhanced viscosity can be formed. This striking phenomenon of structural changes during flow can be even better observed in Figure 2. After a short shearing time, the first normal stress difference rises steeply to a plateau value. The occurrence of normal forces is usually attributed to the presence of finite strains in viscoelastic materials [13]. From this definition follows that only the shear induced state is elastic whereas the solution at rest behaves like a Newtonian liquid. Below the critical shear time of 10 s only viscous properties can be detected:

In analogy to the transient rheological experiments, time dependent flow birefringence measurements can be performed. The birefringence of the solution is given by the difference of the two principal axes of the refraction index ellipsoid n_1 and n_2 [12,14].

$$\Delta n = n_1 - n_2 \tag{2}$$

Some results of these measurements are summarized in Figure 3. We observe a curve similar to that of the first normal stress difference. After the critical shear time of 10 s the flow birefringence increases steeply and after about 20 s a steady-state value is reached. Immediately after the onset of flow, the birefringence Δn does not deviate from zero and under these conditions the solution behaves as a completely anisotropic medium. After a certain shearing time,

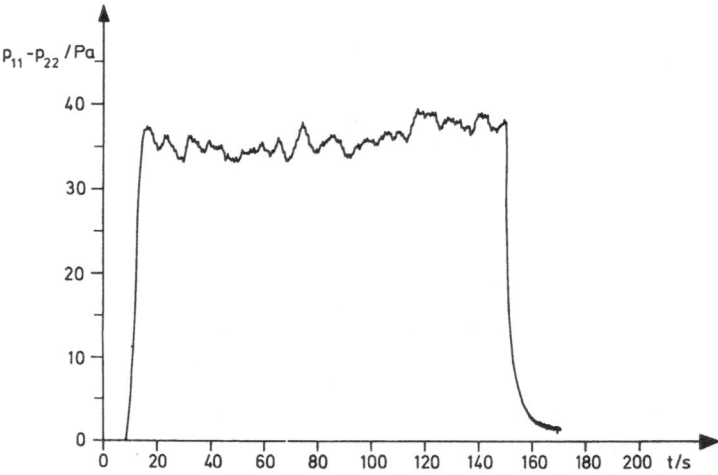

Fig. 2. Dependence of the first normal stress difference $p_{11} - p_{22}$ on time of shearing at $\dot{y} = 200 \, s^{-1}$ (TTAS, $T = 20\,°C$, $c = 2.5 \cdot 10^{-3}$ M)

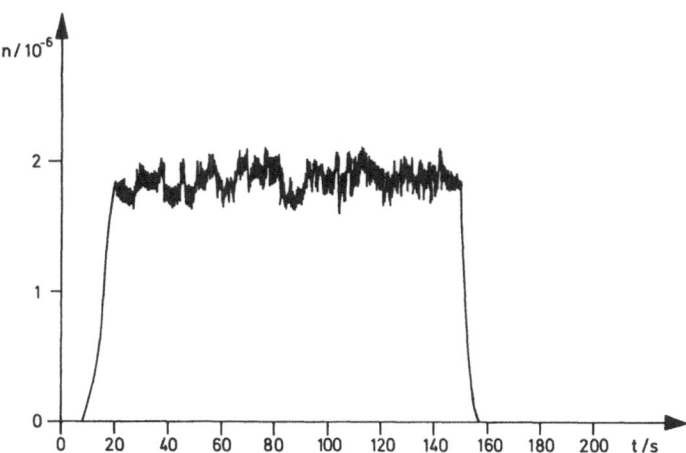

Fig. 3. The flow birefringence Δn as a function of the time of shearing at $\dot{y} = 200$ s^{-1} (TTAS, $T = 20$ °C, $c = 2.5 \cdot 10^{-3}$ M)

however, a flow induced structure can be formed and strong birefringence appears. This phenomenon is directly related to the occurrence of normal stress. The extinction angle χ, which describes the dynamic orientation process of particles under shear conditions, is another important parameter which can be used to characterize the anisotropic flow behavior of solutions. The experimental techniques of flow birefringence allow the determination of this angle χ together with the corresponding birefringence Δn as a function of the shear rate [12]. χ is usually defined as the smallest angle between the direction of flow and one of the two mutually perpendicular extinction positions [12]. The extinction angle χ describes the orientation degree of anisometric particles under flow conditions. Some results of these measurements are summarized in Figure 4.

As soon as the shear induced structure begins to form, the extinction angle χ is zero degrees. That means: the flow induced structure is completely aligned in the streaming solutions. We observe a value of $\chi \approx 0$ already at conditions where the shear induced structure has just formed ($t > 10$ s), which is consistent with the assumption that the supermolecular structure can only grow in the direction of flow. We learn from these experiments that the shear induced phase consists of large, anisometric particles which can easily be aligned in the streaming solution.

When the flow process is suddenly stopped, the shear induced phase decays and the unperturbed state is reforming again. This process can be observed by measuring the time dependent functions $p_{11}(t) - p_{22}(t)$ and $\Delta n(t)$ after cessation of steady-state flow. Some results of this type of measurement are represented in Figure 5.

In this experiment, the shear rate was suddenly reduced to zero at $t = 0$. In the semilogarithmic plot we obtain a straight line describing a monoexponential decay of the shear induced state. The relaxation process after cessation of steady-state flow can obviously be characterized by a single relaxation time constant τ. In this case, we obtain $\tau = 3.4$ s. The data of flow birefringence and the first normal stress difference are in fairly good agreement and from this result we can conclude that there is a simple relationship between these values. The formation of shear induced phases depends on two parameters: the shear time and the shear rate. The influence of flow can be investigated, when the steady-state values of rheological properties are measured at different shear conditions ($t \to \infty$).

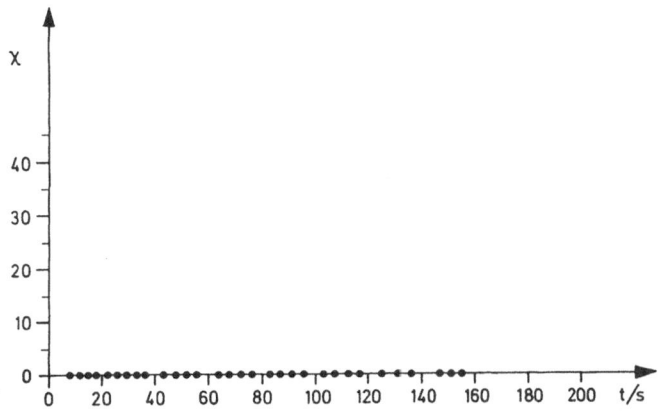

Fig. 4. The extinction angle χ as a function of the time of shearing at $\dot{y} = 200$ s^{-1} (TTAS, $T = 20$ °C, $c = 2.5 \cdot 10^{-3}$ M)

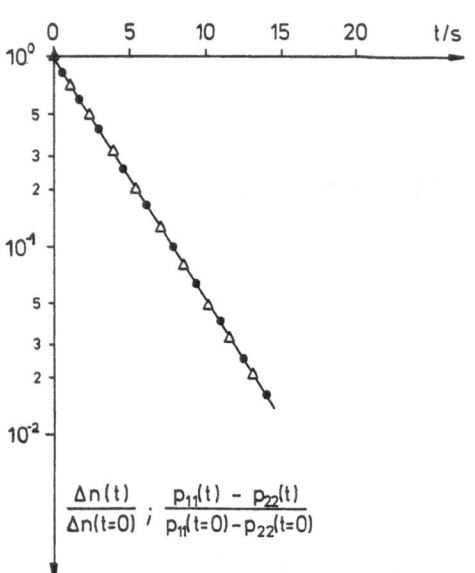

$$\frac{\Delta n(t)}{\Delta n(t=0)} \; ; \; \frac{p_{11}(t) - p_{22}(t)}{p_{11}(t=0) - p_{22}(t=0)}$$

Fig. 5. Decay of the first normal stress difference and relaxation process of flow birefringence after cessation of steady-state shear flow of a solution of $2.5 \cdot 10^{-3}$ M TTAS at $T = 20\,°C$ ($\bullet$: $p_{11} - p_{22}$; $\triangle$: Δn; $\dot{\gamma} = 200\ s^{-1}$)

Some results of these measurements are represented in Figure 6 and Figure 7.

Figure 6 shows the steady-state viscosity $(t \rightarrow \infty)$ as a function of the shear rate and the magnitude of the complex viscosity $|\eta^*|$ as a function of the angular frequency. The complex viscosity can be obtained from dynamic experiments, where a sinusoidal shear rate is applied to the solution.

It can be shown that for most dilute solutions there exists a simple correlation between dynamic and steady-state characteristics [8,15]. For many surfactant solutions the magnitude of the complex viscosity $|\eta^*|$ at a certain angular frequency ω coincides with the steady-state viscosity $\eta(t \rightarrow \infty)$ at the corresponding shear rate $\dot{\gamma}$ [16]. In the case of shear induced phases, however, there is a great discrepancy between these values. The magnitude of the complex viscosity is constant even at high angular frequencies, indicating that the solution still behaves as a simple Newtonian liquid. The experimental results show that flow induced phase transitions cannot occur in the oscillating shear mode. The steady-state viscosity $\eta(t \rightarrow \infty)$, however, points to the existence of a new structure at high rates of shear. The flow induced state can be characterized by an increased viscous resistance. The elastic properties of this phase are given in Figure 7.

Figure 7 shows the first normal stress difference as a function of the shear rate. We observe an increase of about two orders of magnitude at a certain, well defined value of the shear rate. Normal stresses can be understood as forces acting in directions normal to the plane of shear. They are usually attributed to the development of finite elastic strains in the sample. In the regime of low shear rates the solution behaves as a Newtonian liquid and there are no normal stresses different from the hydrostatic pressure.

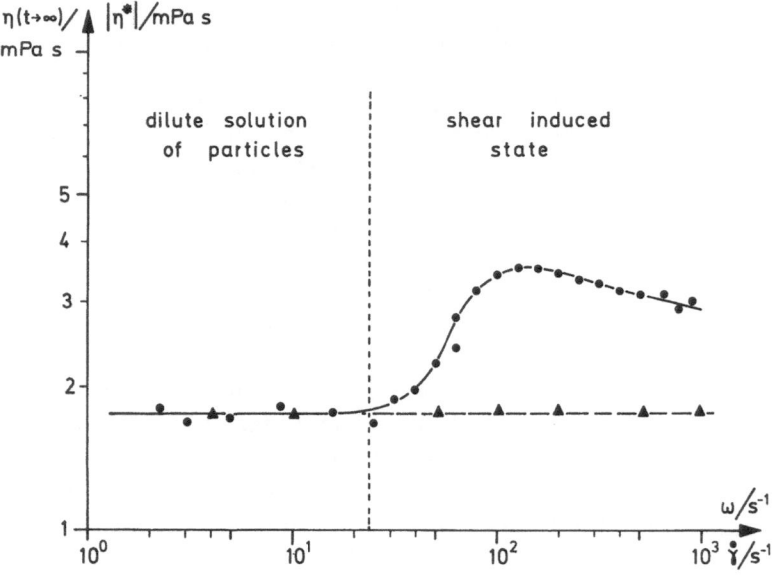

Fig. 6. Dependence of the steady-state viscosity $\eta(t \rightarrow \infty)$ as a function of the shear rate and the magnitude of the complex viscosity $|\eta^*|$ as a function of the angular frequency ω of a solution of $5 \cdot 10^{-3}$ M TTAS at $T = 20\,°C$ ($\bullet$: $\eta(t \rightarrow \infty)$; $\blacktriangle$: $|\eta^*|$)

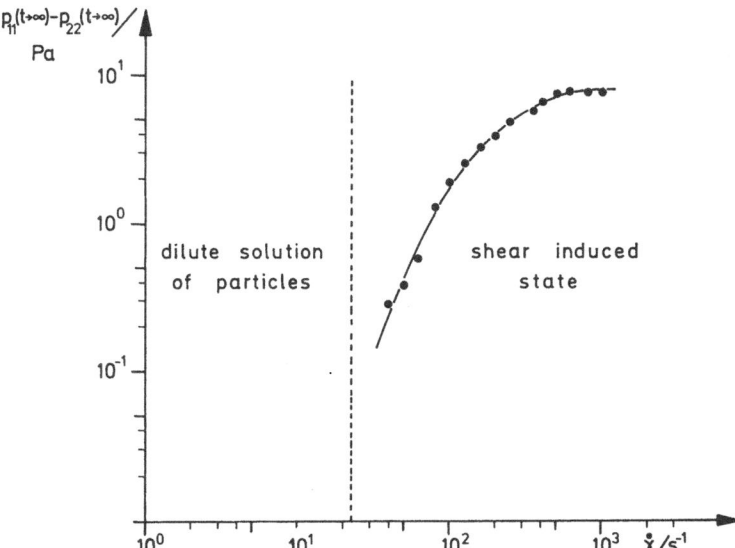

Fig. 7. Dependence of the first normal stress difference $p_{11}(t \to \infty) - p_{22}(t \to \infty)$ on shear rate $\dot{y}$ of a solution of $5 \cdot 10^{-3}$ M TTAS at $T = 20\,°C$

The shear induced phase, however, whose rheological properties are described by complex constitutive equations, shows the appearance of strong normal forces. This phenomenon is attributed to the elastic energy stored in the material during flow. The results show again that elasticity can be induced by shearing of the solutions.

Similar behavior can also be observed from flow birefringence measurements. Relevant results are represented in Figure 8. It is obvious that the curves of normal stress and flow birefringence describe the same physical phenomenon. Both curves point to the existence of a certain well defined threshold value of the shear rate characterizing the phenomenon of flow induced phase transitions. Under conditions of high shear rates we observe flow birefringence of the supermolecular structure. The extinction angle of this phase remains always at a value of about zero degrees. This behavior is represented in Figure 9.

As soon as the shear induced structure can be built up, it is completely aligned in the direction of flow. It is interesting to note that the extinction angle does not

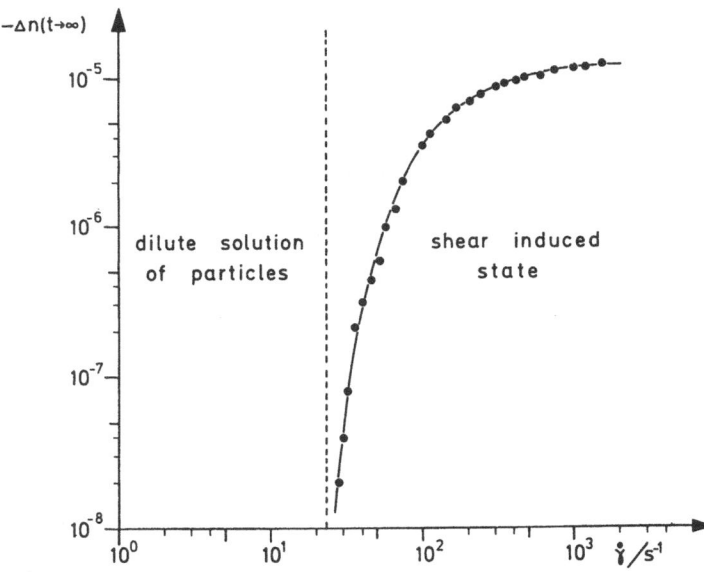

Fig. 8. The flow birefringence $\Delta n(t \to \infty)$ as a function of the shear rate $\dot{y}$ of a solution of $5 \cdot 10^{-3}$ M TTAS at $T = 20\,°C$

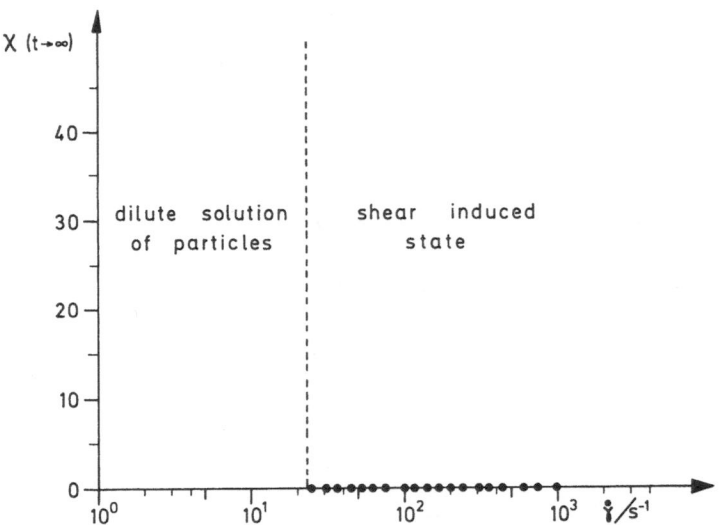

Fig. 9. The extinction angle χ as a function of the shear rate $\dot{y}$ of a solution of $5 \cdot 10^{-3}$ M TTAS at $T = 20\,°C$

depend upon the shear rate although the corresponding flow birefringence and the first normal stress difference show quite different behavior. In order to solve this problem we have to consider the stress optical law which predicts a simple relationship between optical and rheological quantities. Such a law was first proposed by Lodge [17] and also postulated in recent years by Doi and Edwards for solutions containing rigid, overlapping rods [9]. This relationship is based on the idea that flow birefringence can be brought about by stress birefringence according to rubbers and solids, using the same considerations and theories as for the rheological behavior of these systems.

The stress optical law describes a simple linear relationship between the stress tensor and the dielectric tensor ε. This means that the principal axes of the refraction index ellipsoid coincide with those of the stress ellipsoid. The proportionality factor C is called the stress optical coefficient. In polymer systems, it depends upon the anisotropy of the polarizability of a monomer unit of a macromolecule and it has, therefore, a characteristic value for each polymer. From the stress optical law thus defined, the following Equations can be derived [9,12,17]:

$$\Delta n \sin 2\chi = 2Cp_{21} \tag{3}$$

$$\Delta n \cos 2\chi = C(p_{11}-p_{22}) \tag{4}$$

For the shear induced phase we obtain ($\chi = 0$):

$$\Delta n = C(p_{11}-p_{22}) \tag{5}$$

It is clear that for the shear induced phase there is a simple relationship between flow birefringence and first normal stress difference. Thus the validity of the stress optical law can be tested simply by plotting measured values of Δn as a function of the first normal stress difference. Such data are given in Figure 10.

At low flow rates we observe a linear relationship indicating that the stress optical law holds in this regime. From the slope of the curve the stress optical coefficient C can be calculated. For the detergent solution we obtain:

$$C \approx -2 \cdot 10^{-7}\ \mathrm{Pa^{-1}}$$

On the grounds of this simple relationship the relaxation behavior of Δn and $p_{11}-p_{22}$ must be identical. This is clearly evidenced from Figure 5.

The stress optical coefficient C has a characteristic value for each surfactant system. At high rates of shear we observe deviations from the initial linear relationship. Such behavior is predicted by theories which have recently been proposed by Hess and Thurn [18–22] and Doi and Edwards [9]. One important result of these theories is that at high gradients the optical extiction angle does not coincide with the mechanical extinction angle, indicating that the stress tensor is no longer directly proportional to the dielectric tensor [9,18]. In such a regime the simple stress optical law does not hold.

The optical extinction angle χ remains always at zero degrees, although Δn and $p_{11}-p_{22}$ are strongly influenced by the shear rate. This phenomenon points

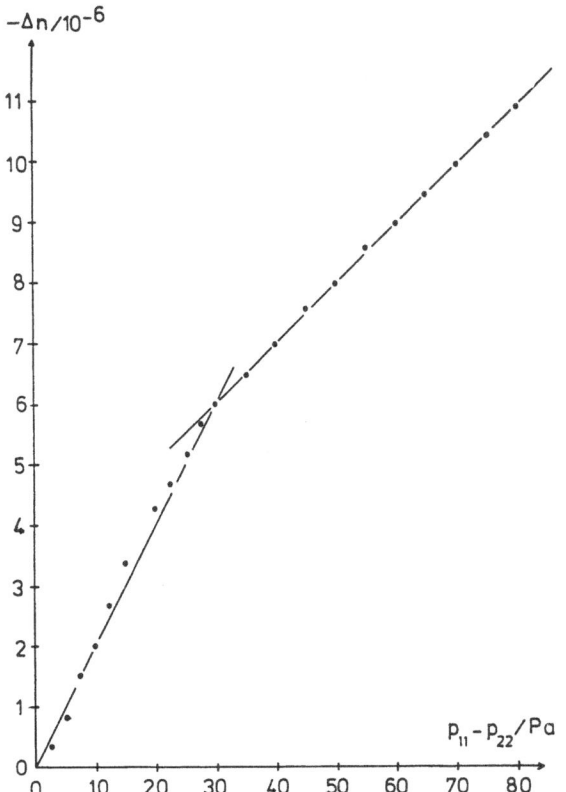

$-\Delta n / 10^{-6}$

$p_{11} - p_{22} / \mathrm{Pa}$

Fig. 10. Verification of the validity of the stress optical law: Δn as a function of the first normal stress $p_{11} - p_{22}$ measured at corresponding shear rates (TTAS, $T = 20\,°C$, $c = 5 \cdot 10^{-3}$ M)

to the existence of flexible particles which can be stretched in the streaming solution or it can be explained by the fact that at high shear rates more material is bound into the flow induced state.

4. Discussion

The stress optical law, which holds for the low shear regime, allows the direct comparison of optical and mechanical properties. From Equations (3) and (4) follows that an extinction angle of about zero degrees must be related to high normal stresses and small shear stresses. For the limiting case of $\chi = 0$ we obtain:

$$p_{21} = \eta_s \cdot \dot{\gamma} \qquad (6)$$

where η_s denotes the solvent viscosity. In this ideal case, the shear induced phase exhibits strong elastic properties without influencing the viscosity of the

solution. The experimental results of Figure 1 and Figure 6 seem to be inconsistent with these predictions. In order to solve this problem, we have to account for the experimental error of χ, which is of the order of 1–2 degrees for the solution being investigated.

From the calculated value of C and the shear stress of the flow induced state a hypothetical extinction angle can be derived from Equation (9). We find $\chi = 0.6$ degrees; a result which does not deviate too much from zero degrees and which is in fact within our experimental error. We can, therefore, conclude that the stress optical law is not only valid for the first normal stress difference, but it also holds for the shear stress (Eq. (3)). From all the available information it seems that the flow behavior of the shear induced phase is strongly influenced by orientation processes. The fact that the super molecular structure is completely aligned in the direction of flow can be understood by postulating the shear induced formation of anisometric aggregates with large axial ratio. The particles which exist in the solutions at rest are too small to be aligned by the applied shear field [1–6]. We must, therefore, conclude that the supermolecular structure is built up from single micelles. The appearance of normal stress must unambiguously be related to the aggregation process of the micelles. If the solution is envisaged to consist of elastic particles which are stretched in the direction of flow, normal forces result. In a rotational flow field between concentric cylinders the concentration force due to be stretched elements acts like a strangulation and forces the solution towards the axis of rotation, resulting in the appearance of normal forces [13]. The molecular origin of this phenomenon is not yet completely understood, but an interesting aspect in the literature is engaged in the normal stress of dilute suspensions of dumb-bells, rigid rods and pearl necklaces [23, 24]. The normal stress of these systems is regarded as a consequence of an aggregation process caused by inelastic collisions between the particles which are induced by the shear field [25]. The first normal stress difference is always due to the presence of anisotropic particles or anisotropic networks because the orientation process is coupled to the entropy elasticity of the system. The stress optical law predicts that the normal stress has the same molecular origin as the flow birefringence. This is also confirmed by the relaxation time constants in Figure 5. The stress optical law holds during the whole relaxation process, indicating that the orientation angle remains at zero degrees. We can, therefore, conclude that the relaxation time τ is not related to the rotational motion of the flow induced particles. This time constant describes the kinetics of phase decay. Our results show that this phase transition follows first order kinetics.

The flow birefringence Δn depends on three independent factors, the optical intrinsic anisotropy, the form birefringence and an orientation factor which is a function of the shear rate and the dimensions of the particles [26, 27]. From measurements of the extinction angle it is possible to calculate the orientation factor but the other two quantities, the form birefringence and the optical intrinsic anisotropy, cannot be calculated from the experimental data. Since it is not possible to separate these two factors a detailed molecular interpretation of flow birefringence cannot be done at the present state. It is, however, possible to present a simple model which accounts for the observed behavior. The results of our measurements indicate that there is a mechanism of coalescence during flow. By shearing it seems possible that the micelles acquire enough energy to overcome the repulsive forces between them. In this context it seems to be of significance that most of the known viscoelastic surfactant systems are built up from cationic surfactants with the pyridinium or trimethylammonium head group and the salycilate counter-ion. From conductivity measurements it is well known that the degree of dissociation is always small, indicating that the counter-ions are strongly bound to the surface of the micellar aggregates. In this way the net charge on the micelles is reduced and electrostatic repulsion has only minor effects.

When the micelles touch they may stick together. They may coalesce and form some kind of supermolecular structure during flow. This shear induced structure can only grow in the direction of flow. With increasing shear rate two different processes can occur. It is either possible that more micelles undergo coalescence or the supermolecular structure can be stretched in the direction of flow. The state of the shear induced structure is evidently complex. It is continuously broken by the shearing forces but also reforms all the time. It is thus characterized by a balance between the destructing and reforming shear forces. After cessation of steady-state shear, the structure decays and the quiescent state is reforming again. The physical changes during shearing are found to be reproducible.

The phenomenon of shear induced phase transitions does not only occur in surfactant systems but can be observed in polymer solutions as well. A large body of experimental data can be found in recent reviews

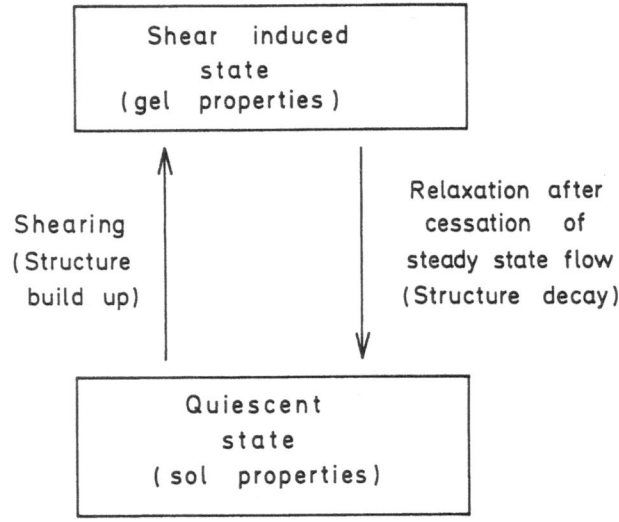

Fig. 11. Schematic drawing of the mechanism of shear induced phase transitions

[28]. P. Bradna and O. Quadrat have started a detailed investigation on polymer solutions of polymethylmethacrylate in different organic solvents [29—31]. As in the case of surfactant solutions they found shear induced phase transitions which depend on shear rate and shear time [30]. From a qualitative point of view, there is excellent agreement between the measurements of Bradna and Quadrat and our own experimental results.

Bradna and Quadrat propose in their paper a semiempirical model which accounts for the described phenomenon [31]. It is based on the process of entanglement formation during flow. In surfactant solutions such network junctions can occur when the micellar aggregates are arranged in the form of long chains like pearl necklaces. These structures are generally assumed to be flexible and hence they may have the same properties as polymer solutions. It is, therefore, possible that the semiempirical model of Bradna and Quadrat can also be applied to surfactant solutions. From a qualitative point of view, we are able to explain all phenomena in terms of such a model.

For surfactant solutions, however, we may have an even more complicated situation because we cannot exclude that the micellar aggregates undergo structural changes during flow. At the present state we do not have enough information to give a precise description of all molecular processes which lead to the formation of the shear induced structure. This problem will be the subject of further investigations.

5. Conclusions

The above observations suggest that shearing of the surfactant solutions can result in reversible formation of a supermolecular structure. The flow induced phase has gel properties, whereas the quiescent state behaves like a simple sol. The system can be used as an element sensitive to shear forces and it can be switched between two completely different states.

Acknowledgement

Financial support of this work by a grant of the „Deutsche Forschungsgemeinschaft (DFG)" is gratefully acknowledged.

References

1. Hoffmann H, Löbl M, Rehage H, (Corso XC) (1985) Physics of Amphiphiles: Micelles, Vesicles and Microemulsions, p 273, and references cited therein
2. Rehage H, Hoffmann H (1982) Rheol Acta 21:561
3. Gravsholt S (1976) J Coll Interf Sci 57:576
4. Gravsholt S (1980) Polym Coll II:405
5. Ulmius J, Wennerström H, Johansson LBA, Lindblom G, Gravsholt S (1979) J Phys Chem 83:2232
6. Hoffmann H, Löbl M, Rehage H, Wunderlich I (1985) Tenside-Detergents, in press
7. Scheraga HA, Backus JK (1951) J Am Chem Soc 73:5108
8. Rehage H, Hoffmann H (1983) Faraday Discuss Chem Soc 76:363
9. Doi M, Edwards SF (1978) J Chem Soc Faraday Trans II 74:418
10. Löbl M (1985) PhD-Thesis, Universität Bayreuth
11. Koeman D, Janeschitz-Kriegl H (1978) Progr Coll & Polym Sci 65:265
12. Janeschitz-Kriegl H, Meißner J (ed) (1983) Polymer Melt Rheology and Flow Birefringence, Springer, Berlin Heidelberg New York
13. Darby R (1976) Albright LF, Maddox RN, McKetta JJ (eds) Viscoelastic Fluids: An Introduction to Their Properties and Behavior, Marcel Dekker Inc, New York, Basel
14. Löbl M, Thurn H, Hoffmann H (1984) Ber Bunsenges Phys Chem 88:1102
15. Cox WP, Merz EH (1958) J Polym Sci 28:619
16. Hoffmann H, Rehage H, Schorr W, Thurn H (1984) Mittal KL, Lindman B (eds) Surfactants in Solution, Vol 1, Plenum Press, New York, p 425
17. Lodge AS (1956) Trans Faraday Soc 52:120
18. Thurn H, Löbl M, Hoffmann H (1985) J Phys Chem 89:517
19. Hess S (1974) Physica 74:277
20. Hess S (1980) Z Naturforsch 35a:915
21. Hess S (1975) Z Naturforsch 30a:728, 1224
22. Hess S (1977) Physica 87a:273
23. Markovitz H (1967) Eirich Fr (ed) Rheology: Theory and Applications, Vol 4. Academic Press, New York, p 347
24. Williams MC (1965) J Chem Phys 42:2988
25. Matsuo T, Pavan A, Peterlin A, Turner DT (1967) J Coll Interf Sci 24:241
26. Peterlin A, Stuart HA (1939) Z Phys 112:129
27. Peterlin A, Stuart HA (1939) Z Phys 113:663
28. Rangel-Nafaile C, Metzner AB, Wissbrun KF (1984) Macromolecules 17:1187
29. Quadrat O (1972) Collection Czechoslov Chem Commun 37:980
30. Bradna P, Quadrat O (1980) Coll & Polym Sci 258:626
31. Bradna P, Quadrat O (1984) Coll & Polym Sci 262:189

Received December 3, 1985;
accepted May 20, 1986

Authors' address:

H. Hoffmann, H. Rehage, Ingrid Wunderlich
Institut für Physikalische Chemie
der Universität Bayreuth
Universitätsstraße 30
D-8580 Bayreuth, F.R.G.

Progress in Colloid & Polymer Science   Progr Colloid & Polymer Sci 72:60–82 (1986)

The non-homogeneous thermodynamically autonomous and equivalent microphase*)

H.-G. Kilian

Abteilung Experimentelle Physik, Universität Ulm, F.R.G

Abstract: By defining non-homogeneous mixed extended chain crystals as autonomous and equivalent microphases, a thermodynamics of eutectoid multi-component systems is formulated through which the superstructure will be linked to the molecular conditions. The general consequences of this approach will be elucidated by a description of crystallization and melting in *n*-alkane multi-component systems and in radiation cross-linked polyethylene networks.

Key words: *n*-alkane mixtures, networks, thermodynamics, eutectoid multi-component systems, autonomous and equivalent microphase.

Introduction

Attempts to describe crystallization and melting of copolymers in terms of classical thermodynamics have encountered substantial difficulties [1–7]. In the first place this arises from the fact that "non-homogeneous extended-chain-sequence" lamellae of finite size are formed on crystallization. These lamellae cannot be described without breaking away from classical concepts.

Because of showing extended-chain crystallization, it suggests itself to study multi-component oligomer systems so that one does not have to deal with folded-chain crystallization and "crystal network constraints". This was the motivation for starting a comprehensive examination of *n*-alkane multi-component systems and extended-chain crystallizing fractions of linear polyethylene [8–14]. With the chain-ends always segregated into longitudinal defect-boundaries, "non-homogeneous" mixed extended chain lamellae are formed. To consider these lamellae as thermodynamically autonomous and equivalent micro-phases is equivalent to developing the thermodynamics of eutectoid oligomer multi-component systems where the thickness distribution of the microphases is clearly

determined by the molecular conditions (molecular-weight distribution).

Slight modifications were necessary only for applying the same theory to crystallization and melting of networks [15–18].

Oligomers

To underline the particular role of the chain-ends we use the representative model of an *n*-paraffin chain which is drawn in Figure 1.

In order to shed light on the central issue regarding the nature of crystallization-induced chain-end segre-

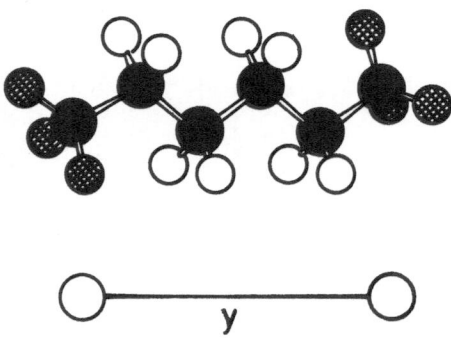

Fig. 1. Hexan molecule and its abstract model

---

*) Lecture presented during the 32nd Annual Meeting of the Kolloid-Gesellschaft, Berlin October 2–4, 1985.

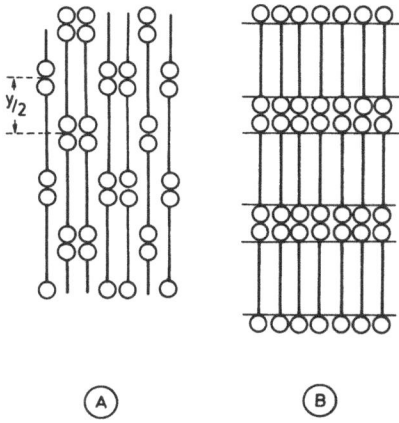

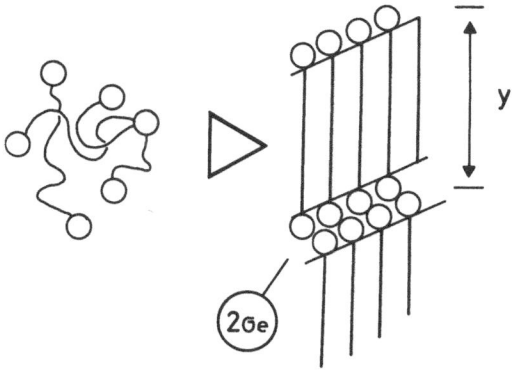

Fig. 2. (A) Scheme of a nematic *n*-alkane single crystal allowing for random nematic „lattice-compatible" steps y/2 only. (B) Scheme of the crystalline state of lowest energy where the chain-ends are segregated into crystallographic layers

Fig. 3. Illustration of the chain-end segregation which is required for forming *n*-alkane single-component crystals. The thickness of the sub-lamellae is uniquely related to the chain length parameter y, the number of chain-standing carbon atoms. $2\sigma_e$ is the molar excess-free-enthalpy per chain

gation let us consider the "nematic crystal" in Figure 2. Always having the chain-ends in couples, superchains of infinite lengths are formed.

Because Platzwechsel of the "superchains" are not possible, nematic structures like that in Figure 2 A could only be realized as "static configurations" within an energy-minimum lattice.

The energy-excess situation in both of the examples shown in Figure 2 is consequently of interest. Let us define the nematic excess enthalpy per chain-end pair within the lattice by

$$\Delta H_{ex,n} = 2\sigma. \tag{1}$$

To form chain-extended lamellae, the chain-end pairs must apparently behave like lattice incompatible chemical units which on crystallization become segregated into chain-end double-layers (see also Figure 3). Hence; the excess-energy of a chain-end pair within the lattice should substantially exceed that associated with adjacent chain-ends at the crystal surface [21]. We are therefore led to the inequality

$$2\sigma \geqslant 2\sigma_e \tag{2}$$

where $2\sigma_e$ is the interfacial excess-enthalpy within the well-ordered stacks of lamellae, a two-dimensional sketch of which is depicted in Figure 2 B.

Theoretical calculations of the defect-energies of a 2g1-kink within the polyethylene lattice [22] support the idea that

$$2\sigma \geqslant 1.6 \text{ kcal/mol/CH} \tag{3}$$

always holds true whereby the concrete value $2\sigma_e$ corresponds to the molar surface enthalpy as derived from the description of the melting temperatures of even *n*-paraffin homologues [8–14].

According to the above considerations the smallest nematic motions as observed for *n*-paraffins at elevated temperatures [31–33] become possible only if longitudinal chain motions become increasingly decoupled. Such effects are clearly indicated with the high temperature broadening of the longitudinal defect layers [13].

Polyoxymethylene oligomers [4, 21] and various other flexible chain systems [4, 9, 23, 24] seem to form extended-chain lamellae as well. Chain-end migration represents the normal crystallization behaviour for linear, stereoregular and flexible chains.

Chain-end pairs apparently behave like excessive chemical lattice defects. It is therefore easy to generalize the above ideas by defining "non-crystallizable chemical units" in general terms. This is indeed the key step which leads from the treatment of oligomer multi-component systems to an analogous thermodynamical description of "sequence-extended crystallization" in eutectoid polymers.

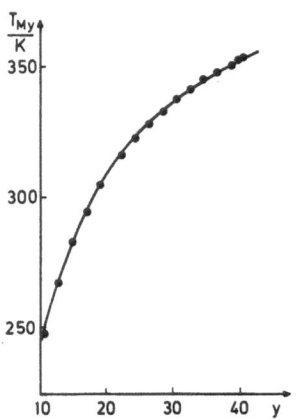

Fig. 4. Melting temperatures of even n-alkanes, T_{My}, versus the chain-length parameter, y. The solid line is computed with the aid of Equation (4) using the parameters [8, 12]: $\Delta h(T) = (700 - 1.4 (418.2 - T))$ cal/mole CH_2 $2\sigma_e(T)/\Delta h(T) = 2.9828$

$\Delta h(T)$ and $\sigma_e(T)$ are the molar melting enthalpy and the molar interfacial free enthalpy per unit the temperature dependence of which are defined by [6–12, 25]

$$\Delta h(T) = \Delta h(T_M) - \Delta C(T_M - T) \tag{6}$$

$$\sigma_e(T) = \sigma_e(T_M) \, \Delta h(T)/\Delta h(T_M) \tag{7}$$

whereby ΔC is the difference of heat capacities. y is the number of carbon units such that the interfacial free enthalpy $\sigma_e(T)$ describes the excess free enthalpy of the chain ends related to bulk free enthalpy of the CH_2 units.

The thermodynamic parameters per chain unit [6–13] which were used in the calculations, were found not to depend upon the length of the chains. We are thus led to conclude:

The interfacial and the intrinsic properties of a single lamella cannot depend to a measurable extent upon its thickness.

The excess properties within the double-chain-end layers seem to be determined by pair-interactions with a surprisingly small radius of influence.

Hence, it appears to be allowed to consider each of the extended-chain lamellae as a "fully crystallized autonomous microphase". *It is important to realize that in a single component system all of these extended-chain lamellae are identical such that they are altogether thermodynamically equivalent.* In contact with the melt each of these lamellae has for example the same melting temperature, T_{My}. Melting of the total ensemble of identical lamellae occurs in a zipper-like manner at constant intensive variables (T_M, p) such as to produce jump-like changes of extensive variables (see Fig. 5). Each of these coexisting ensembles of equivalent crystal-microphases is indeed in the state of an indifferent phase equilibrium. Hence, the system displays exactly the phenomena of a phase-transition as derived for a single component system by the traditional thermodynamics [19, 20].

From a mathematical point of view, to formulate the equivalence conditions for each of the microphases is to *reduce the number of independent variables to the well known set of macroscopical variables necessary for the complete description of the thermodynamical properties of a single component system.*

At first sight, the micro-phase model appears rather artificial. Yet, one has to be aware that it clearly shows how segregation of "non-lattice-compatible" chemical units (for pure n-alkane systems the chain-ends that are "non-compatible with the CH_2-lattice) can be

Accordingly, we find ourselves in the position to formulate the very important hypothesis:

On crystallization "lattice-incompatible" chemical units or "lattice incompatible defects of stereo-regularity" should always be segregated so as to form lamellae as superstructure elements. The thickness-distribution of extended-chain or extended-sequence crystals should therefore be uniquely related to the configuration of the "non-lattice-compatible chemical units" (chain ends, short-chain branchings or other co-units).

It is one of the objects of this paper to prove the utility and the consequences of this hypothesis.

Oligomer single-component systems [4, 8–14, 21–36]

Because of always observing n-alkane single crystals comprising stacks of extended-chain lamellae, the chain-ends behave like "non-crystal-lattice compatible units" (see Fig. 3).

The melting temperatures of homologue n-alkanes with an even number of carbon atoms can be fairly well described with the aid of the Flory-Vrij equation [8, 25] (Fig. 4)

$$T_{My} = T_M \left(1 - 2\sigma_e/\Delta h \, y\right)/(1 + (RT/\Delta h \, y)\ln y) . \tag{4}$$

T_M is the maximum melting temperature

$$\lim_{y \to \infty} T_{My} = T_M . \tag{5}$$

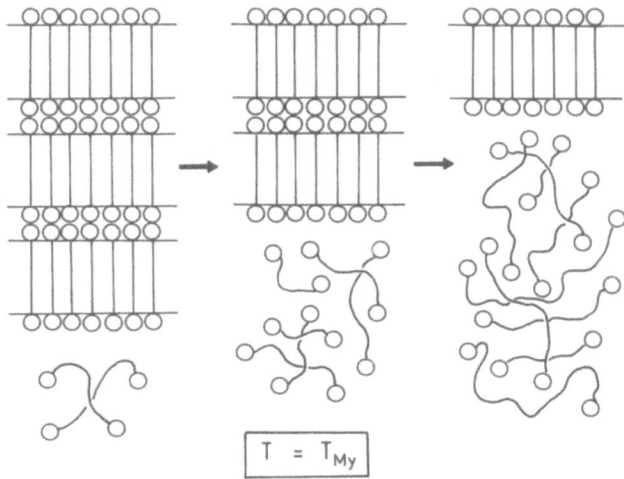

Fig. 5a. Illustration of the consecutive melting within a cluster of identical thermodynamically autonomous and equivalent extended-chain lamellae

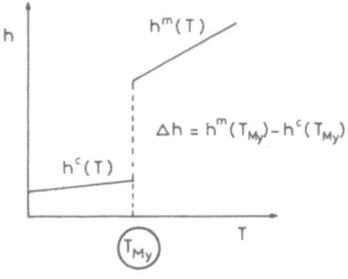

Fig. 5b. The "jump-like" change of the enthalpy at the melting process of an assembly of identical micro-phases which is formed on extended-chain crystallization in oligomer single component systems

introduced into the thermodynamics. Largely as a consequence of this circumstance, substantial hope was engendered that this model should be particularly adequate in application to crystallizing multi-component *n*-alkane systems. Moreover, the convenience of application of our model assures its use for correlating conditions of the chemical structure and relevant parameters of the superstructure in eutectoid multi-component systems in which no folded-chain or folded-sequence crystallization occurs.

Mixtures of *n*-alkane homologues

Binary solid mixtures [4, 8, 10–14, 39–41]

As an extension of the previous notation we let y_i represent the chain length of the components (in the present section $i = 1, 2$ with $y_2 > y_1$). Extended-chain lamellae must be formed under all circumstances. The scheme in Figure 6 illustrates the new situation: If mixed extended-chain lamellae are formed, the atomistic order within the longitudinal layers is necessarily distorted due to the mismatch of chain lengths $\Delta y_{12} = y_2 - y_1 > 0$. This is in accord with a SAXS pattern analysis which reveals lowered average electron-densities within the longitudinal layers [8, 13, 14]. Within the boundaries the averaged density decreases continuously, as is illustrated in Figure 7. The "density profile" represents an equilibrium property of the microphase which can no longer be treated as homogeneous. Hence, the non-homogeneous microphase appears to

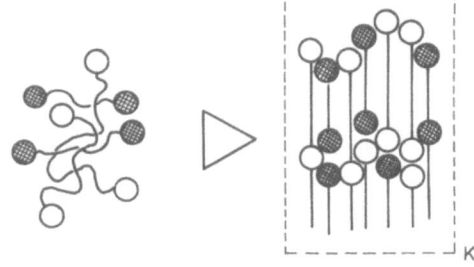

Fig. 6. Model of a mixed extended-chain lamella with distorted longitudinal layers. K_{12} indicates the existence of clusters comprising a large number of regularly staggered lamellae of identical thickness

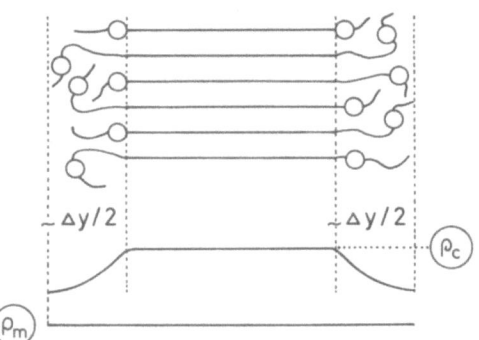

Fig. 7. Model of a binary extended-chain non-homogeneous microphase. The average density within the longitudinal boundaries falls continuously below the value within the crystal lattice, yet always remains larger than the density in the melt. The course of the density function can be considered to characterize the "non-homogeneity" within the crystallized microphase

be a model just cut out for bringing off a solid description of mixed extended-chain lamellae.

There is, nevertheless, coherent X-ray scattering within larger lamellae clusters [14] (K 12 in Fig. 6).

The lateral dimensions of the lamellae mostly become exceedingly large, as can be seen by evidence from the electron-micrograph shown in Figure 8. To have established thermodynamical equilibrium, it is necessary to have strictly the same composition within each of the lamellae. The well defined and highly ordered superstructure may thus simply be considered to be the consequence of having identical and very broad lamellae densely packed together.

To have the chains always located either within the solid or the liquid mixed phase, is of fundamental importance because of having the solid and the liquid mixed microphases submitted to the same conditions of coexistence as given by the traditional thermodynamics.

For the binary non-homogeneous microphase as cooperative unit, the molar free enthalpy may be defined by

$$g = n_1 \, y_1 \, g_{01} + n_2 \, y_2 \, g_{02} + 2 \, \sigma_e$$
$$+ \, RT \, (n_1 \ln (x_1) + n_2 \ln (1 - x_1))$$
$$+ \, A_{12} \, x_1 \, (1 - x_1) \, (n_1 + n_2) \tag{8}$$

Fig. 8. Electron micrograph of a replica which shows stacks of chain-extended mixed crystals formed within a binary *n*-alkane mixture comprising 50 percent in weight of each of the components

where g_{0i} is the molar free enthalpy per unit of the single component system at the same T, p as in the solid mixture. n_i gives the molar numbers of the components, while x_i is the molar fraction of the component i

$$x_i = n_i/n; \; n = n_1 + n_2; \; x_1 + x_2 = 1. \tag{9}$$

Presuming each of the chains y_i to have the same cross-section within the solid solution, we are allowed to define the mixing entropy in ideal terms, thus, being simply related to the molar fractions x_i [8, 10, 19, 20, 38]. By expecting that the excess properties within the boundaries of the non-homogeneous microphase are governed by pair-interactions, hope is engendered that the defect-characterization can to a fist approximation be met by a molecular interpretation of the second virial coefficient, A_{12}, only.

The conditions of coexistence read

$$T^{(c)} = T^{(m)}$$

$$p^{(c)} = p^{(m)}$$

$$\mu_i^{(c)} = \mu_i^{(m)}$$

$$\mu_i^{(c)} = \mu_{io}^{(c)} + RT \ln (x_i^{(c)}) + A_{12}^{(c)} \, (1 - x_i^{(c)})^2$$

$$\mu_i^{(m)} = \mu_{io}^{(m)} \, RT \, (\ln (\varphi_i^{(m)}) + 1 - y/\langle y \rangle^{(m)})$$
$$+ \, A_{12}^{(m)} \, (1 - \varphi_i^{(m)})^2 \tag{10}$$

where $\mu_i^{(\alpha)}$ is the chemical potential of the component i within the phase (α) $((m)$ and (c) characterizing the melt- or the crystal-phase) related to the standard potentials $\mu_{io}^{(\alpha)}$. The chemical potential in the melt is described by means of the Flory-Huggins model [8, 38, 42–47] using the volume fractions φ_i.

$$\varphi_i^{(m)} = n_i^{(m)} \, y_i \, / \sum_k n_k^{(m)} \, y_k. \tag{11}$$

The entropy of mixing is related to the average chain length within the binary melt

$$1/\langle y \rangle^{(m)} = \sum_i \varphi_i^{(m)}/y_i. \tag{12}$$

We once more want to stress that the above coexistence conditions are related to the assembly of solid non-homogeneous equivalent microphases coexisting with the mixed melt. All of the interfacial excess properties should be accounted for by defining the average properties within the longitudinal boundaries of the crystallized lamellae. Hence, if there are molten regions of finite size they should also be considered to represent thermodynamically autonomous and equivalent microphases. Due to the defined number of equivalence conditions the original set of variables is in any case reduced to the classical number of independent macroscopical variables. This is a very important fact for guaranteeing that mass fraction and composition of each of the sets of the coexisting non-homogeneous equivalent microphases are uniquely determined at each temperature in exact accordance with the classical lever rule.

Under the above circumstances, the stability of the solid mixture should depend upon quality and concentration of defects within the longitudinal boundaries which are themselves not uniformly distributed over the total microphase volume. This might be one reason for simplifying the phenomenological characterization due to having exceedingly small excess entropy effects. Defining consequently a second virial-coefficient which does not depend on temperature, a binary solid microphase can be formed only if the "traditional condition of stability" [19, 20]

$$T_c = A_{12}/(2R) > T \tag{13}$$

is not violated.

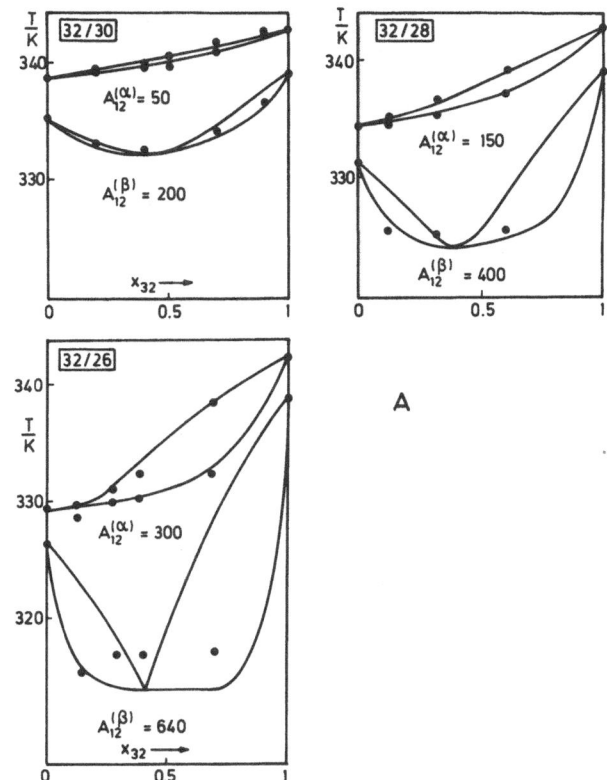

A

Fig. 9 A. Plot of isobaric state diagrams of binary n-alkane systems comprised of $k_2 = 32$ and $y_1 = 30, 28, 26$ according to [8, 12]. The solid lines have been computed using the parameters given in Figure 4. The second virial-coefficients $A_{12}^{(i)}$ have been assigned to the values indicated within the homogeneous regions in the solid state. The $A_{12}^{(i)}$ are seen to increase with increasing disparities in chain-length

Comparison with experiments

It is now very satisfactory that even isobaric binary state diagrams of n-alkanes can be calculated by only adjusting the excess parameter A_{12} (see Fig. 9) [8, 10–13].

When the liquidus is reached on cooling, crystallization starts by forming mixed extended-chain lamellae accompanied by a defined fractionation of the components. Mass-fraction and composition of the coexisting phases, the binary melt und the binary "crystals", have continuously to be changed on further cooling so as to always establish the binary phase-equilibrium (see Fig. 9 B):

On passing the two-phase region mass-fraction and composition of both of the coexisting phases are constantly transformed for thermodynamic reasons.

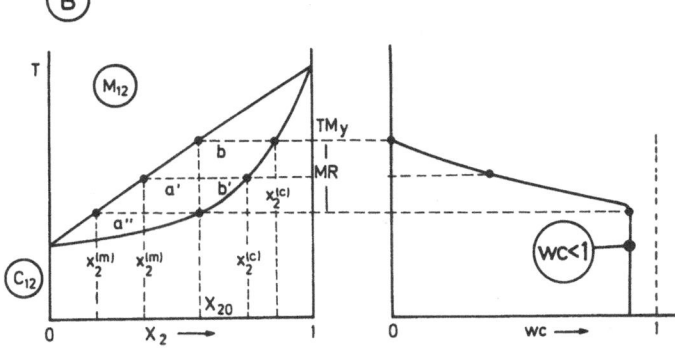

Fig. 9 B. Schematic drawing of the binary isobaric state diagram with binary mixed crystals. On cooling crystallization starts when the liquidus is met. The particles become fractionated ($x_2^{(c)} > x_2^{(m)}$). Within the heterogeneous region, mass fraction and composition of the coexisting phases is uniquely determined according to the lever rule. The area of the heterogeneous region increases with the interaction parameter A_{12}. In any case, the melting process always has to cover a temperature range as illustrated with the $wc - T$ plot drawn at the right hand of the schematic drawing

According to the analysis of synchroton-SAXS pattern an example of which is shown in Figure 10, the superstructure should be characterized by a lamellae cluster comprising more than six or eight densely packed, coherently scattering non-homogeneous microphases. The number of "visible" oders of interferences is seen to become smaller at elevated temperatures (see Fig. 10). Apparently, the width of the defective layers increases due to rising nematic motions of the chains [31, 33, 48]. This exceptionally large chain mobility within the microphases at elevated temperatures should as such guarantee that internal equilibrium is rapidly approached.

This conclusion appears to be defended by the finding that the excess parameter, A_{ik}, is clearly related to the relative disparity of chain-length [8, 10–13]

$$A_{ik} = A \Delta y_{ik}/\langle y_{ik}\rangle; \; A = \text{const}; \; \Delta y_{ik} = |y_i - y_k| \quad (14)$$

where $\langle y_{ik}\rangle$ defines

$$\langle y_{ik}\rangle = (y_i + y_k)/2. \quad (15)$$

This result underlines that the equilibrium-excess energies within the boundaries should be controlled by pair-interactions, thus showing no dependence on the lamella thickness.

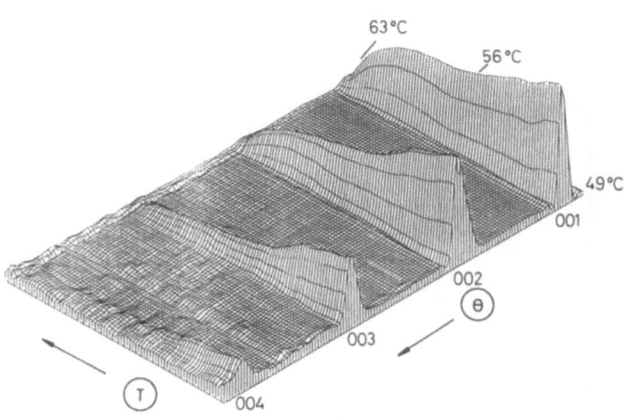

Fig. 10. Synchrotron-SAXS pattern against the temperature for the binary system comprising .5 weight-percentage of each component with $y_1 = 28$ and $y_2 = 32$ (the Bragg angle). The intensities are corrected so as to represent the total power scattered by the sample. The decreasing intensities within a set of intereferences must therefore be due to increased average widths of the longitudinal defect layers. These effects are produced by temperature stimulated decoupled nematic motions of the chains at elevated temperatures

On heating each of the SAXS patterns has been taken within 20 seconds [13]. Three or four orders of (001)-interferences could be identified in each case. Anticipating later results we are led to the conclusion: Since crystallization segregation has to comply with the lever rule, chain diffusion has to proceed under conditions which permit that clusters comprising more than eight identical lamellae are formed within a period of time shorter than 20 seconds.

The equivalent subvolumina

If "interfacial energies" are approximately independent upon the microstructure that is developed on crystallization, we come to the very significant conclusion:

Crystallization segregation of the components could be confined to "equivalent subvolumina". These subvolumina correspond to elementary bricks of any superstructure developed on crystallization. Their equivalence even on crystallization rests squarely on the fact of having autonomous microphases the formation of which allows to always adjust "internal equilibrium" within each of these subvolumina.

Binary eutectic mixtures [4, 8, 10–13, 39]

The disparity of chain length in the solid mixtures becoming larger than $\Delta y_{12}/\langle y_{12}\rangle > 0.2 - 0.3//$, a "mixing gap" in the solid state is predicted [8, 10–13]. If we have

$$T_C^{(c)} = A \Delta y_{12}/\langle y_{12}\rangle/2R \geqslant T_{M\,\text{max}} \quad (16)$$

where $T_{M\,\text{max}}$ is the highest melting temperature of the single component systems involved, the mixing gap covers de facto the total concentration range. Pure extended-chain crystals (comprising chains of uniform length) can be formed only. Two different segregation processes have now to occur on eutectic crystallization

— "lamella segregation" of the chain-ends

— "crystallization segregation" of the chains y_i so as to form single-component-extended-chain lamellae

These processes are illustrated by the scheme in Figure 11.

The liquids of eutectic binary state diagrams of n-alkanes can be nicely fitted as can be seen by evidence from Figure 12. It should be mentioned that excess energies in the binary melt must be accounted for. The molar enthalpies of mixing computed from this are in good agreement with calorimetric data [49] (see Fig. 13). It turns out that the stability of binary

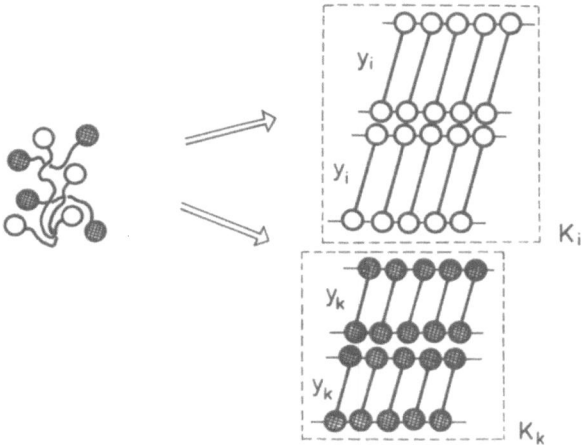

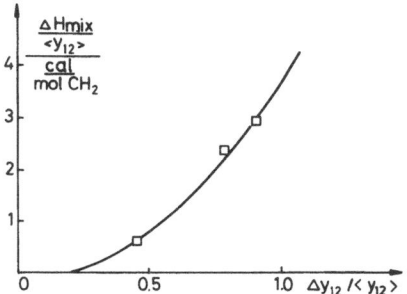

Fig. 13. The solid line depicts the molar mixing enthalpy per unit in binary *n*-alkane melts dependent on the relative disparity in chain lengths $\Delta y_{12}/\langle y_{12}\rangle$ ($y_1 = 10$ mixed with $y_2 = 12, 16, 20, 24, 28$) according to [8,12]. The open squares are data calorimetrically measured by van der Waals and Hermans [49]

Fig. 11. Illustration of eutectic segregation of chains and chain ends by forming clusters K_i each of them composed of identical lamellae y_i

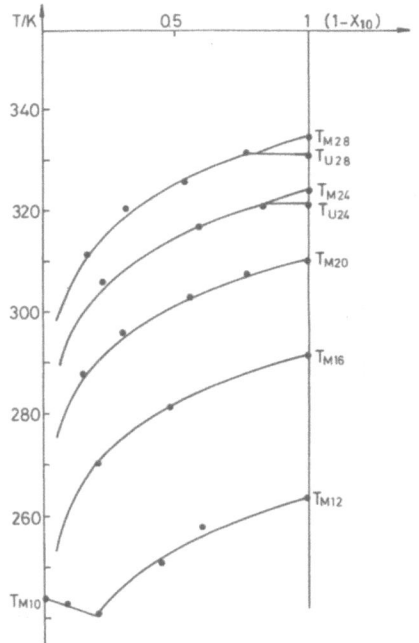

Fig. 12. Plot of isobaric state diagrams of binary eutectic *n*-alkane systems, $y_1 = 10$ mixed with $y_2 = 12, 16, 20, 24, 28$ according to [8,12]. The solid lines are calculated

ic state diagram rapidly degenerates, at least showing only the liquidus of the longest chains (see Fig. 12). The sequence of crystallization is then fixed within the total range of compositions so as to always let the longer chains crystallize first.

Let us illustrate crystallization within a binary eutectic system with a concrete example. On cooling a binary melt of the composition x_{20} (see Fig. 14), the

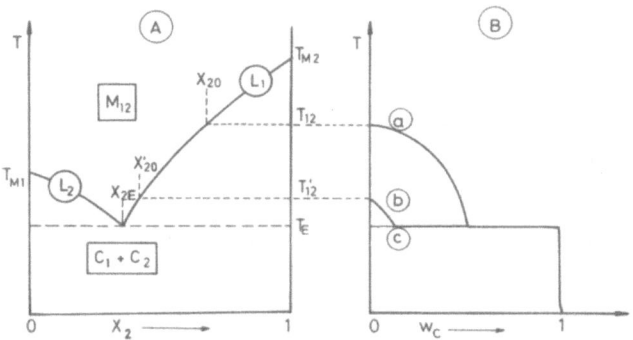

Fig. 14. Isobaric eutectic state diagram. T_{Mi} is the melting temperatures of the single component system i, T_E the eutectic temperature. x_{20} and x'_{20} are the overall molar fractions showing the melting points T_{12} and T'_{12} on the liquidus $L1$. In the given examples, crystallization starts with single component crystals (1). On cooling, the massfraction of the crystals (1) increases as depicted in terms of the degree of crystallinity w_C in part B of the figure. The bending is uniquely related to the thermodynamic properties in the binary melt M_{12} so that a calculation of the liquidus is equivalent to determining the excess quantities in the melt. When the single component crystals (2) are also formed at the eutectic temperature, simultaneous crystallization of both crystals C_1 and C_2 has to occur such as to let the eutectic composition within the binary melt unaltered (indifferent phase equilibrium according to the phase rule). Hence, a jump-like change of extensive variables (like the molar enthalpy) is the typical phenomenon of eutectic crystallization in binary systems. This is, of course, to be seen by the topological features of the degree of crystallinity-plot as shown in part B of this figure

n-alkane melts is limited to a maximum disparity in chain length.

The asymmetry of the state diagrams is a consequence of the Flory-Huggins model [8, 38, 42–47] due to the mixing entropy in the melt being uniquely related to the relative disparity of chain lengths. The eutect-

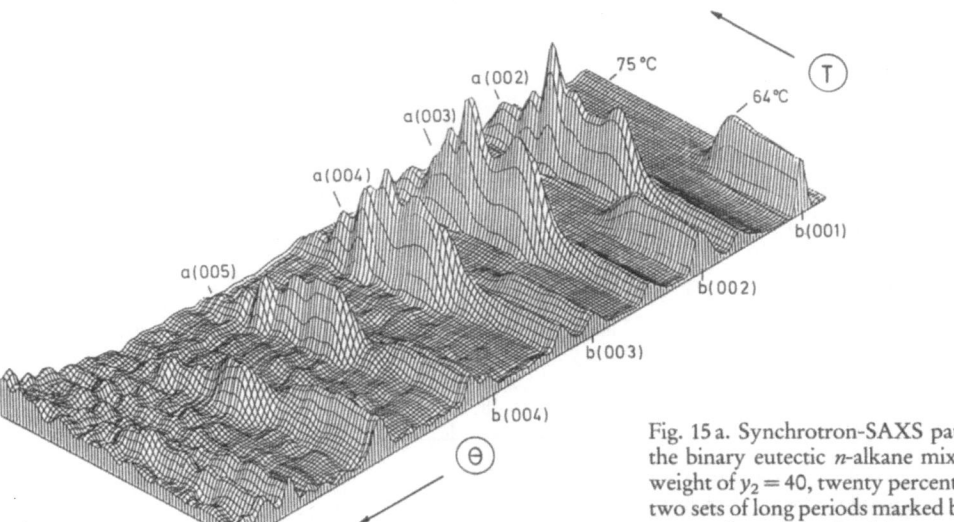

Fig. 15 a. Synchrotron-SAXS pattern against the temperature for the binary eutectic n-alkane mixture comprised of 80 percent in weight of $y_2 = 40$, twenty percent of $y_1 = 30$ according to [13]. The two sets of long periods marked by a and b give evidence that clusters are formed, each of them comprising identical lamellae only

first extended-chain lamellae of the longer chains appear when the liquidus is met. At each temperature $T_E < T < T_{Mi}$, the mass fraction of these crystals is determined corresponding to the lever rule.

It is pertinent to eutectical systems that the pure crystals, once formed in the two-phase region, stay in principle unaltered. On cooling crystallization occurs by growth or by nucleation of new crystals.

T_E is the eutectical temperature. According to the phase rule we find here simultaneous formation of a mixture of pure crystals y_1 and y_2 regulating the mass fractions of the coexisting phases so that the eutectical concentration in the melt remains unaltered. According to the phase rule for a binary system three coexisting phases are in a state of indifferent phase equilibrium so that extensive variables have to change "jump-like" at constant intensive coordinates (see Fig. 14 B).

The synchrotron-SAXS patterns display in general two different long periods, thus proving that well ordered clusters K1 and K2 are formed, each of them comprising at least six identical lamellae (see Fig. 11). The eutectical mixture of pure crystal lamellae C_1 and C_2 is often characterized by the appearence of small clusters where lamellae of different thickness may be built into a single cluster [13]. Some of these features can be picked up from the electron-micrograph as shown in Figure 16.

It is possible to compute the heat exchange in the melting range of eutectical systems as measured with a calorimeter [8, 10–12] with the aid of our thermodynamical equations. Such calculations are successful

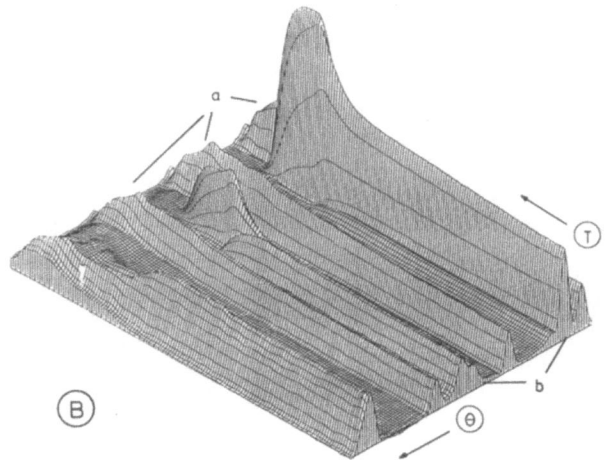

Fig. 15 b. Synchrotron-XAXS pattern against the temperature for the binary eutectic n-alkane mixture comprised of 70 percent in weight of $y_2 = 40$, 30 percent of $y_1 = 30$ according to [13]. A few degrees below the eutectic temperature, a part of the single-component crystals (1) ($y_1 = 30$) undergo a transformation of the lattice modification [13] such that two long periods are left. The lattice transformation apparently seems to cover stacks comprising more than six or eight identical lamellae

independet of the systems composition (Fig. 17). Within the experimental accuracy, the macroscopical response seems not to depend to a measurable extent on the superstructure which is different for differently composed eutectical systems. *It is worth remarking that such behavior should be found if each of the extended-chain lamellae y_2 and y_1 largely behave like thermodynamically autonomous and equivalent microphases.*

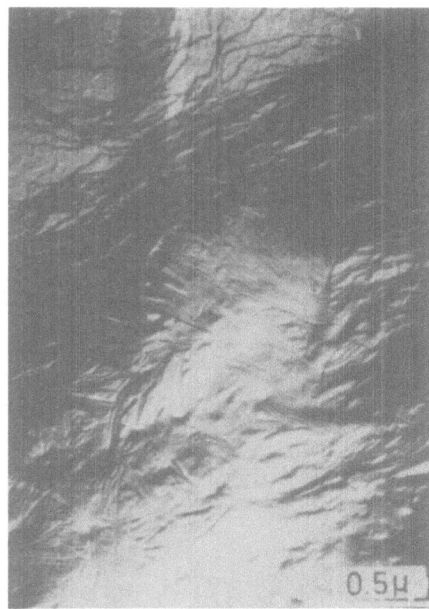

Fig. 16. Electron micrograph of a replica which shows the superstructure of a binary eutectic *n*-alkane system comprising 80 percent in weight of $y_2 = 40$, 20 percent of $y_1 = 30$ according to [13]. The eutectic mixture of crystals C_1 and C_2 is to be seen at the bottom of the picture

Topological characteristics of the superstructure

Having extended-chain crystallization only, the thickness of the "primary elements of the superstructure", the non-homogeneous microphases, is clearly determined by the "chain-length situation". Well ordered clusters comprising identical lamellae seem to represent elementary bricks of the superstructure. How these bricks are finally arranged so as to produce the light-microscopical picture is not yet known at all. One interesting aspect should nevertheless be mentioned, namely, that eutectic binary systems mostly display a more or less pronounced and extensive superstructure that topologically differs from more granular pictures for binary systems with mixed extended-chain crystals (Fig. 18).

A reason behind that difference might be given by the following argumentation: in eutectic systems pure crystals are formed which are always stable. The architecture of the superstructure should thus have causal connections with the history of the crystallization segregation. In contrast, solid solutions have continuously to adjust their composition and their mass fraction according to the conditions of phase equilibrium. The superstructure might therefore finally be organized in a manner different to that in eutectical systems.

Multicomponent oligomer systems

The preceding considerations embody the generalization to multi-component systems composed of oligomer homologues. It clearly emerges that limited solubility in the solid solutions should determine the degree of crystallization induced molecular-weight segregation in multi-component systems.

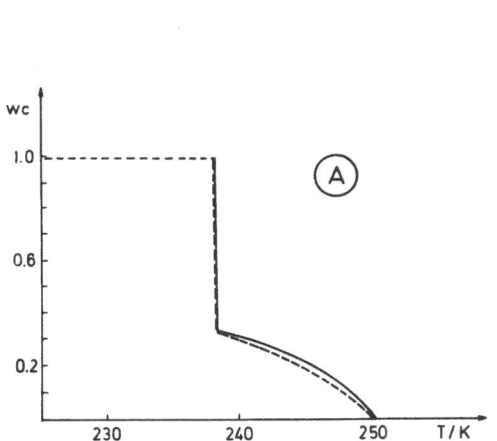

Fig. 17 A. Plot of the molar degree of crystallinity, w_C, against the temperature for an eutectic binary *n*-alkane mixture comprised of chains $y_1 = 10$ and $y_2 = 12$ with the volume fraction $\varphi_{10} = .496$ according to [8, 11]. The dotted line is calculated

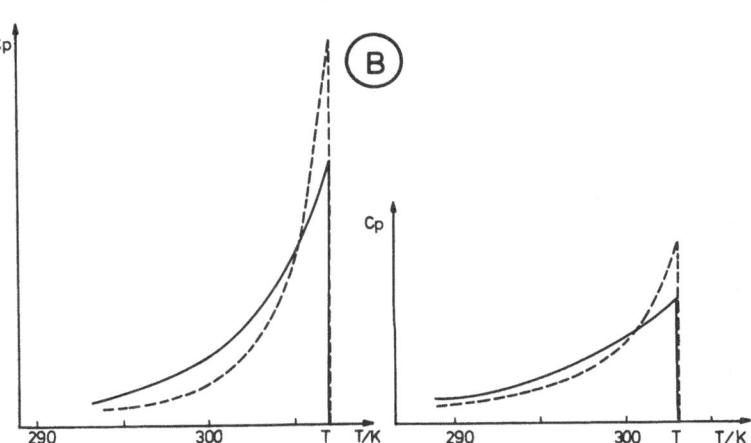

Fig. 17 B. $cp(T)$ curves of eutectic binary *n*-alkane mixtures comprised of chains of the lengths $y_1 = 10$ and $y_2 = 20$ with molar fractions of $x_{10} = .76$ (diagram on the left hand) and $x_{10} = .55$ (on the right hand). The experiments are de-smeared, the dotted lines are calculated [8, 12]

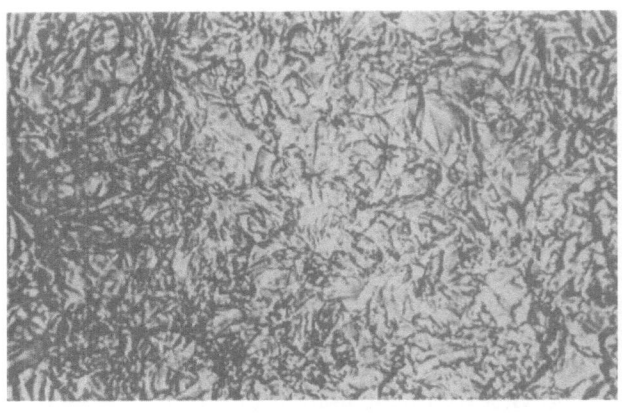 

a)   b)

Fig. 18. Light-microscopic pictures of binary n-alkane systems (a) with chain-extended mixed crystals (b) as eutectic system

To accomodate our understanding to the growing problems we take advantage of learning gradually by increasing the total number of components step by step. Based on these results and their description, we then focus our discussion on solid solutions comprising many components. The central question here is whether and under which circumstances extended-chain mixed crystals become defect-saturated.

Ternary n-alkane mixtures

Treating the ternary melt as an ideal athermic mixture and relying on the assumption that the defect situation is controlled by pair interactions, the chemical potential can straigthforwardly be formulated [19, 20, 38, 50]

$$\mu_i^{(c)} = \mu_{io}^{(c)} + RT \ln \left(x_i^{(c)}\right) + g_{ex\,i}^{(c)} \tag{17}$$

whereby the partial molar excess-enthalpies should simply result from additive pair interaction terms, each one characterized by the second virial coefficient as obtained from the investigation of binary systems

$$g_{ex\,1}^{(c)} = A_{12}^{(c)} x_2^{(c)} \left(1 - x_1^{(c)}\right) + A_{13}^{(c)} x_3^{(c)} \left(1 - x_1^{(c)}\right)$$
$$- A_{23}^{(c)} x_2^{(c)} x_3^{(c)}$$

$$g_{ex\,2}^{(c)} = A_{12}^{(c)} x_1^{(c)} \left(1 - x_2^{(c)}\right) - A_{13}^{(c)} x_1^{(c)} x_1^{(c)}$$
$$+ A_{23}^{(c)} x_3^{(c)} \left(1 - x_2^{(c)}\right)$$

$$g_{ex\,3}^{(c)} = A_{12}^{(c)} x_1^{(c)} x_2^{(c)} + A_{13}^{(c)} x_1^{(c)} \left(1 - x_3^{(c)}\right)$$
$$+ A_{23}^{(c)} x_2^{(c)} \left(1 - x_3^{(c)}\right). \tag{18}$$

It is now very satisfactory that the experimental "liquidus line" of isothermal cuts applied to a ternary n- alkane system with ternary extended-chain crystals, can be fitted perfectly using the "binary $A_{ik}^{(c)} \Leftrightarrow A_{ik}$" as defined in Equation (14) (Fig. 19). Quasi-binary cuts for an eutectoid ternary system with binary mixed crystals are depicted in Figure 20. The data calculated with the aid of the binary A_{ik} again show surprisingly excellent correspondence to the experimental results [8, 10]. The quality of the representation of the total melting process can be seen by evidence from the plot in Figure 21.

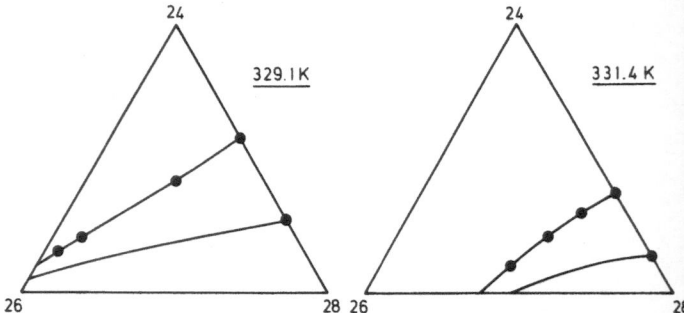

Fig. 19. Isobaric isothermal cuts of the Gibbs representation of the state diagram of ternary n-alkane system showing ternary extended-chain mixed crystals comprising chains of the lengths $y_1 = 24$, $y_2 = 26$ and $y_3 = 28$ at two different temperatures as indicated with each of the diagrams according to [8, 12]

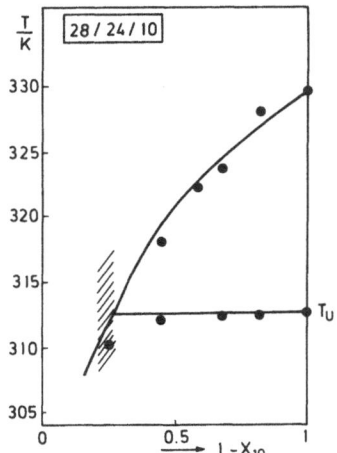

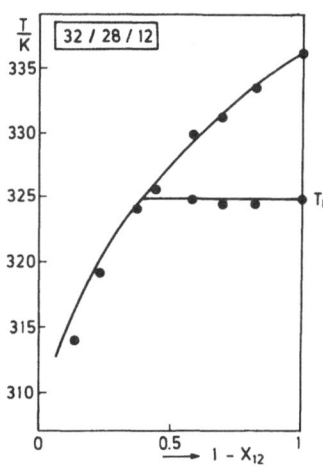

Fig. 20. Plot of the liquidus obtained by quasi-binary cuts of the Gibbs representation of ternary *n*-alkane mixtures comprised of components the y_i of which are indicated with each of the drawings according to [8,12]. In these circumstances binary extended-mixed crystals can only appear. The molar fractions of the components were kept constant such that the concentration of the shortest chains is continuously increased as indicated on the abcissa of the plots. The concentration of the chains which enter into the mixed crystals are for $x_{10} = 0$ on the left: $x_{24} = x_{28}$, on the right: $x_{28} = .37, x_{32} = .73$. The solid lines are theoretical, the dashed area indicates the region where a solid-solid phase seperation seems to occur because of coming below the binodal

This indeed defends the idea that the "lattice field effects" within stacks of identical lamellae seem to be as small as to justify application of the model of thermo-dynamically autonomous and equivalent non-homogeneous microphases. The defect situation within the longitudinal boundaries seems cooperatively be related to the equilibrium conditions within the total microphase. Within the boundaries next-neighbor interactions should be dominant having their radius of influence squeezed down to very short distances.

n-alkane multi-component mixtures [4, 8, 11–13]

To further prove the utility of our concepts, let us now focus on the question of how we are enabled to predict the sequence of crystallization in multi-component *n*-alkane mixtures with the aid of the thermodynamics. To keep the discussion as clear as possible, let us simply indicate the temperature where the different types of mixed microphases crystallize for the first time [13]. It must indeed be taken as a severe proof of our model as to how typical "crystallization patterns" of isobaric multi-component state diagrams (fully represented by more or less complicated hyperfaces) are correctly computed for different systems always using the "pair interaction excess parameter" A_{ik} as defined with Equation (14).

Representative results are compiled in Figure 22 while their synchrotron-SAXS pattern are shown in Figure 23. It can be seen that the typical sequence of crystallization segregation of chains is theoretically fairly well understood, of course within the limits of accuracy. Here, it might be necessary to correctly

account for the excess properties in the multi-component melt. In the present case, the melt has always been treated as an ideal athermic mixture.

For all of the *n*-alkane multi-component systems investigated, clusters have been observed which are comprised of at least six identical lamellae (see Fig. 23).

That substantial rearrangements might occur within crystallized *n*-alkane multi-component systems, is clearly manifested by the splitting of the long period which is shown by the synchrotron-SAXS pattern depicted in Figure 24. Below the temperature of $T = 317$ K (as indicated by an arrow in Fig. 24) the ternary mixed extended-chain crystal is no longer stable so that the total number of coexisting microphases is increased by forming two new and stable solid solu-

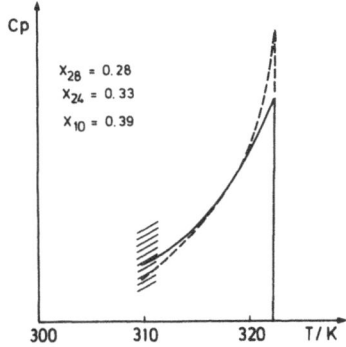

Fig. 21. Plot of $dwc/dT \sim cp$ in arbitrary units against the temperature for a ternary *n*-alkane system comprised of chains of the lengths $y_1 = 10$, $y_2 = 24$ and $y_3 = 28$ with the concentrations as indicated with the figure. The dotted line is theoretical [8,12]. The solid line is obtained by de-smearing calorimetric measurements [8,12]

Figs. 22 and 23. Theoretical and experimental results in *n*-alkane multi-component systems. Chain length parameters y_{ik} and compositions are: (A) y_{ik}: 20/34/46/40, φ_{ik}: 25/25/25/25, A = 4000; (B) y_{ik}: 20/22/28/32/40, φ_{ik}: 20/20/20/20/20, A = 8000; (C) y_{ik}: 20/22/28/32/40/44, φ_{ik}: 8/11/13/19/23/26, A = 8000

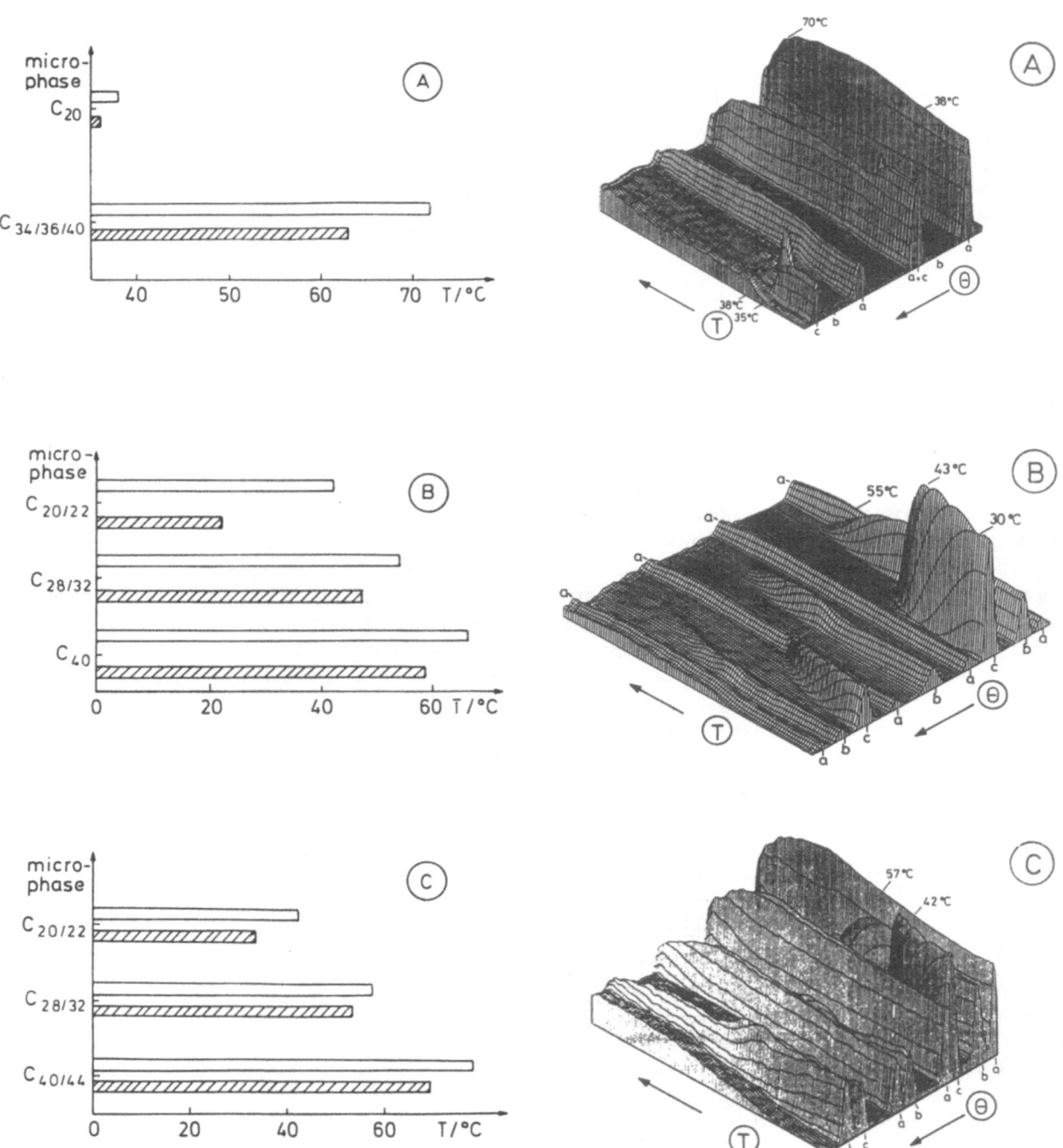

Fig. 22. Plots on the left: Observed temperatures at which crystals comprising chains as indicated with the indices appear for the first time and the theoretical predictions according to [13]. (▨▨▨ theoretical)

Fig. 23. Drawings on the right: Synchrotron SAXS pattern against the temperature showing a set of (00L)-interferences for each of the microphases under discussion. Hence, there is evidence that coherent clusters are found to be always formed in any case comprising more than six identical lamellae [13]

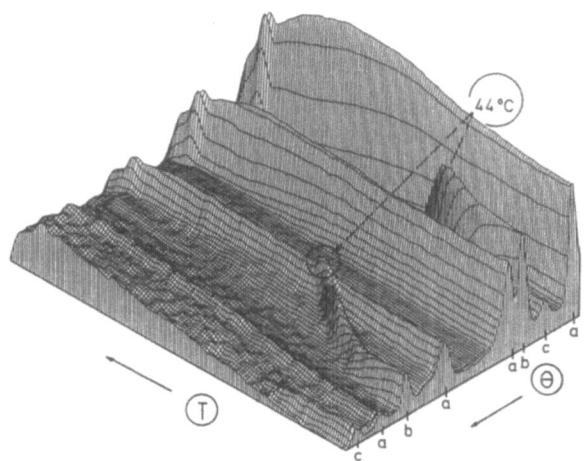

Fig. 24. Synchrotron-SAXS pattern of an *n*-alkane five-component system against the temperature according to [13]. On cooling phase separation starts at the temperature around 44 °C as indicated with the figure. Two new long periods are born out of a single interference. It is evidenced at least by the long period which changes strongly over the temperature that the concentration within the microphases is continuously changed in accordance to the demands of the thermodynamics. If four peaks of different order are observed, more than six identical lamellae must be packed into each of the coherent clusters. Hence, segregation of chains of different lengths must basically be finished within the cluster ensemble in periods of time shorter than 20 seconds, the experimental time scale within which a single SAXS pattern was measured

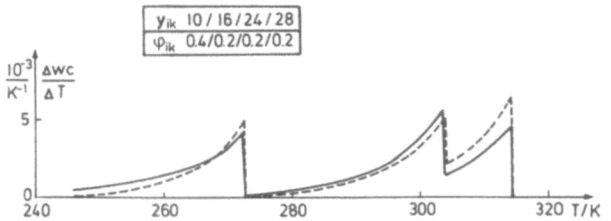

Fig. 25. Plot of dwc/dT against the temperature for an *n*-alkane four-component system comprised of chains with the lengths and with a volume fraction as indicated with the drawing according to [8,11]. The dotted line is theoretical, the solid line represents the de-smeared experiment (DTA measurements)

tions. On further cooling, the degree of splitting increases, thus clearly indicating that mass fraction and composition of these new microphases must continuously become adjusted in accordance with the multi-microphase equilibrium conditions. What is striking is that the above process runs as "solid-solid" phase transition (solid-solid demixing process). Even in these circumstances "six-lamellae clusters" are formed in periods of time less than 20 seconds.

It is of real significance now that independent of the particular features of the superstructure, we always found ourselves in the position to predict the sequence of crystallization. Through this it is also ensured that the total melting process is known. This is once more documented by the example given in Figure 25.

We are again led to the conclusion that all of the extended-chain crystals should operate on melting like autonomous and equivalent, non-homogeneous microphases with intrinsic equilibrium properties that do not depend on the superstructure characteristics. In these circumstances, all of the extensive quantities like the systems enthalpy should display a melting process

where composition and mass-fraction of the coexisting sets of equivalent microphases obey the "lever rule translated into the hyper-space of the isobaric state diagram of a eutectoid multi-component system".

Our model thus delivers a consistent understanding of the macroscopical melting phenomena in multi-component *n*-alkane systems for which otherwise a case is not easily made out. Beyond that, the model allows us to uniquely correlate essential parameters of the primary superstructure and the "molecular conditions" within mixtures of homologues.

The problem of defect saturation

The problem we want to deal with in this section is directed to the question whether a maximum disparity in chain lengths might appear, produced by configurational constraints imposed upon protruding chain segments. To pose this question is equivalent to discussing whether mixed extended-chain lamellae at least should show a limited solubility due to defect saturation.

From the thermodynamical description of the melting of extended-chain crystallized polyethylene fractions [8,10,11,13] it is suggested that the maximum disparity in chain length is given by

$$\Delta y_{max}/\langle y \rangle = M. \tag{19}$$

This seems to hold true irrespective of the type of the molecular-weight distribution. It is now a question of importance whether the above relationship is consistent with the empirical formulation as given in Equation (14).

At first, let us simply ask which "critical disparity in chain length" is predicted for allowing the critical temperature of a binary mixed crystal to just exceed the highest single-component melting temperature (system y_2). This condition is expressed by

$$T_C = A \, \Delta y_{12}/\langle y_{12}\rangle/2R = T_{M2}; \; A_{ik} = A \cdot \Delta y_{12}/\langle y_{12}\rangle \, . \tag{20}$$

Because solid mixtures are discussed in this section, extra "phase symbols" are no longer attached. The calculated data depicted in Table 1 prove that the critical relative disparity in chain length" is nearly invariant for all of the binary mixtures. The results are in accord with the differences in chain lengths beyond which even n-alkane mixtures are found to show eutectic crystallization [4, 8, 10–13, 27, 29, 39].

While these calculations are satisfactory in predicting the molecular conditions for eutectical crystallization in binary systems, the results for more-component, even for ternary systems are discrepant. By adding the excess pair interaction terms as given with Equation (18) and by the use of relation (14), the mixing entropy is obtained to increase so much as to overcompensate the energy-excess terms. Mixing gaps are therefore shifted to lower temperatures. The consequence is that the "critical maximum disparity of chain-lengths" grows to such large values that they become extremely inconsistent with Equation (19).

Let us therefore modify our model in the following manner (see sketch in Fig. 26): Protruding chain segments are assumed to occupy a certain volume the size of which depends on the averaged conformation of the segments involved. The number of distinguishable configurations should be diminished due to steric exclusion effects. This can be calculated by defining quasi-permanent complexes "Co". The cross-section which is covered by each of these complexes may be described by

$$y_2^{co} = 1 + \gamma \, (\Delta y_{12}/2) \tag{21}$$

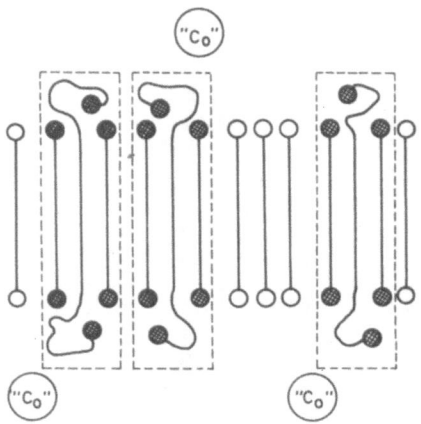

Fig. 26. Model of a binary extended-chain mixed crystal wherein the longer chains are assumed to occupy an increased cross-section within the longitudinal boundaries. It is this larger cross-section by which the longer chains become excluded from some lattice sites. The loss of configurational abilities can be described by considering the "complexes" which are indicated by the dotted rectangles as "quasi-permanent particles. A correspondingly decreased number of particles per unit volume is essential for computing the mixing entropy

letting y characterize its averaged size. The molar number of the short chains is therefore reduced due to the loss brought about by the chains which are "bound" to a complex

$$n_1^{co} = n_1 - \gamma(\Delta y_{12}/2) \, n_2 \, . \tag{22}$$

The critical temperature T_C^{co} is derived to be equal to

$$T_C^{co} = (2A^{co}/R) \, \Delta y_{12}/\langle y_{12}\rangle/(1 + (1/(1 + \Delta y_{12}/2)^{1/2})^2 \, . \tag{23}$$

With the aid of this relation it is also possible to correctly predict for binary systems what the size of the critical disparity of chain length is beyond which one always observes eutectical crystallization (see Table 2).

Table 1. "Critical eutetical" disparity in chain-length

y_2	y_2	Δy_{12}	$\Delta y_{12}/\langle y_{12}\rangle$	T_{my_2}/K	T_C/K[a]
10	12	2	.18	240	273
20	26	6	.26	300	328
40	52	12	.26	350	352
100	130	30	.26	380	391
200	262	62	.27	400	403
300	396	96	.28	408	414

[a]) $A = 1500$ cal/mole, Equation (20)

Table 2. $n1 = n2 = n3 = 100$

y_1	y_2	Δy_{12}	$\Delta y_{12}/\langle y_{12}\rangle$	$w_i/\%$[a]
10	14	4	.18	.82
20	26	6	.26	.74
40	50	10	.22	.78
100	122	22	.2	.8
200	240	40	.18	.8

[a]) w_i "internal degree of order"

The consequences of the model become primarily relevant in multi-component extended-chain lamellae. This can be demonstrated by computing the mixing entropy for ternary extended-chain crystals. We come to the equations

$$- \Delta S_{\text{mix}}^{(co)} / R = \sum \varphi_i^{(co)} / y_i \ln (\varphi_i^{(co)}) \qquad (24)$$

where

$$x_i^{(co)} = n_i^{(co)} / n^{(co)}; \quad n^{(co)} = \sum_{i=1}^{N} n_i^{(co)} \qquad (25)$$

with

$$n_1^{(co)} = n_1 - (y/2) (\Delta y_{12} n_2 + \Delta y_{13} n_3)$$

$$n_2^{(co)} = n_2 - (y/2) \Delta y_{23} n_3$$

$$n_3^{(co)} = n_3 . \qquad (26)$$

The volume fractions $\varphi_i^{(co)}$ are derived to be given by

$$\varphi_i^{(co)} = n_1^{(co)} / n_\Delta$$

$$\varphi_2^{(co)} = n_2^{(co)} (\Delta y_{12} (y/2) + 1) / n_\Delta$$

$$\varphi_3^{(co)} = 1 - \varphi_1^{(co)} - \varphi_2^{(co)} \qquad (27)$$

where

$$n_y = n_1^{(co)} + n_2^{(co)} y_2 + n_3^{(co)} y_3$$

$$y_2 = y/2 \, \Delta y_2; \quad y_3 = y/2 \, (\Delta y_{13} + \Delta y_{23}) . \qquad (28)$$

We learn from the theoretical data complied in Table 3 that the mixing entropy of ternary mixtures decreases when the maximum chain length only is raised.

Hence, it seems to be very likely that the mixing entropy within multi-component extended-chain crystals is diminished due to handicaps enforced by the protruding segments of longer chains. The consequence is that for thermodynamic reasons a maximum density of defects is only tolerated within each of the extended-chain lamellae characterized by a maximum relative chain-length difference.

Problems that might be seen behind the above considerations should serve as a reminder that the thermodynamic description presented, though generally reliable in predicting the eutectoid crystallization characteristics, should be completed by a discussion of the

Table 3. Ternary systems

$y_1 = 20; \ y_2 = 20; \ y/2 = 0.1$ $(- \Delta S_{\text{mix}}/RT)_{\text{ideal}} = 1.099$	
y_3	$(- \Delta S_{\text{mix}}/RT)_{\text{MAX}}$
24	0.936
26	0.899
28	874
30	855
50	0.787
$(- \Delta S_{\text{min}}/RT)_{\text{MAX}}$ (binary $y_1 = y_2$) = 0.693	

concrete defect situation within a non-homogeneous microphase. It is necessary in the first place to provide appropriate means, both experimental and theoretical, for elucidating the molecular structur within the boundaries and for subjecting them to quantitative characterization.

Fractions of polyethylene as eutectoid multi-component systems with chain-extended mixed crystals [4, 8, 11–13, 51, 53]

To describe the eutectoid crystallization of n-alkane homologues with a chain length distribution (Fig. 27) we utilize the above considerations by postulating that the maximum disparity of chain length within the non-homogeneous microphases should be given by [8, 10, 11, 13]

$$\Delta y = M \langle y \rangle; \quad M = \text{const} \qquad (29)$$

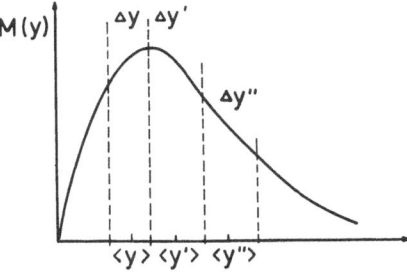

Fig. 27. Schematic illustration of the relationship between the molecular-weight distribution and the thickness distribution of the extended-chain mixed lamellae the solubility of which is limited. Only chains within the range of Δy can enter into a mixed crystal of the average thickness $\langle y \rangle$

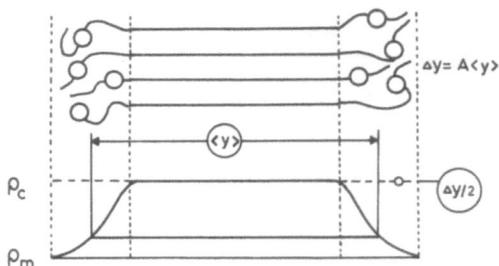

Fig. 28. Model of an extended-chain mixed crystal with the maximum disparity in chain-length Δy. The average chain length within the microphase is y. In the outer parts of the distorted layers the density is assumed to fall to the same value as in the melt, ϱ_m (ϱ_c: density within the crystal core)

Fig. 29. Electron micrograph of a replica of an internal fracture surface of a bulk crystallized fraction of polyethylene according to Anderson [54]

with $\langle y \rangle$ as the average chain length with the mixed chain-extended crystal. Under this assumption we are led to the very universal relationship

$$\phi(y) \mid_{y_{\min}} \overset{\geqq}{\underset{\leqq}{=}} K(y_c) : y \geq y_{\min} \tag{30}$$

correlating the molecular-weight distribution $\phi(y)$ and the crystal-thickness distribution function $K(y_c)$ in the range beyond the minimum length of crystallized chains ($y_{\min}$). The type of the molecular weight distribution may be of any nature whatever, provided that folded chain crystallization does not occur.

With a broad chain length distribution the coexisting microphases are brought into the state of defect saturation: In these circumstances, it is possible to bring the maximum number of different chains into the extended-chain crystal which is defined by Equation (26). The "state of disorder" within the outer parts of defect layers should now approximate the liquid-like structure as in the melt (see sketch in Fig. 28). Because of having then cut out any "crystal field correlation" between neighbored lamellae, *defect-saturated non-homogeneous microphases should in fact be thermodynamically autonomous in accordance with its thermodynamically strict interpretation.*

It is desirable to examine carefully whether the above model holds true. Foremost of all approximations involved is the idea that no "contact excess energy" between neigbored lamellae should appear. A statistically random spatial distribution of the microphases is then expected.

From an analysis of the electron-micrograph depicted in Figure 29 [54] lamellae varying in thickness are found to be nearly randomly scattered. Their

average thickness and even their thickness distribution are in good accord with the knowledge of the size of the average molecular weight and the molecular-weight distribution of the fraction investigated [8, 54]. Moreover, SAXS patterns of fractions of polyethylene with extended-chain crystals show at least only one broad peak (Fig. 30). With the aid of Hosemann's linear theory of paracrystals [55] it was possible to fit calculations to the experimental data under the assumption that the constituents are randomly distributed over the total lamella-cluster ensemble [56].

Hence, examination of the extended-chain lamella superstructure in fractions of linear polyethylene defends the assumption of defect saturated mixed microphases, the thickness distribution of which strictly correlates with the molecular-weight distribution.

To arrive ultimately at an analytical formulation of the melting process in fractions with extended-chain crystals we adopt the following approximations [8, 10, 11]:

— Coexistence only of the fraction with the extended-chain lamellae of smallest thickness and the multicomponent melt will be considered.

— We confine our attention to chains the length of which corresponds to the average chain length in the microphase of smallest thickness.

— The solubility within the extended-chain lamellae is a priori defined by the empirical relation (29).

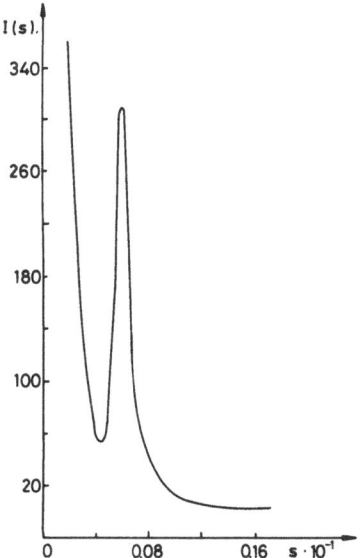

Fig. 30. SAXS pattern of a fraction of linear polyethylene with an average chain length of $\langle y \rangle = 142$ measured at room temperature [56]

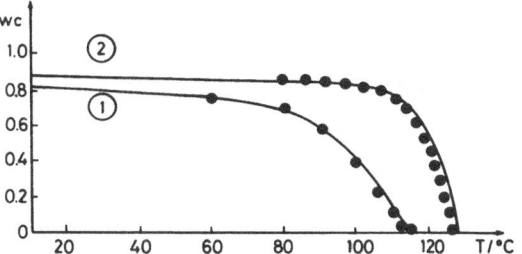

Fig. 31. The molar degree of crystallinity against the temperature for fractions of linear polyethylene with different average chain lengths ((1) $\langle y \rangle = 206$; (2) $\langle y \rangle = 232$) according to [8, 59]

— It is assumed that the total ensemble of mixed extended-chain crystals with $y^* > y$ is "quasipermanent" so that the relative fraction of crystals can be computed from the knowledge of the average thickness, $y(T_{My})$, of the smallest microphase which is just stable at the given temperature T_{My}.

At the coexistence temperature T_{My}, the average thickness of the extended-chain crystal of smallest thickness, $y(T_{My})$, which is just stable at this temperature, is well defined according to

$$T_{My} = T_M \left(1 - 2\,\sigma_e/((1 - M/3)\,\Delta h\,y)/N1\right) \tag{31}$$

$$N1 = 1 + (RT/((1 - M/3)\,\Delta h\,y)\,(\ln(y/(y/2 + 1)$$
$$- \ln(a_y^{(m)})/x_y^{(c)}) + (y - 1)\,x/2) \tag{32}$$

where $a_y^{(m)}$ is "particle-size activity" of the chains y in the melt defined by

$$a_y^{(m)} = x_y^{(m)} \exp\left(1 - y/\langle y \rangle^{(m)}\right). \tag{33}$$

$x^{(m)}$ is the excess parameter per chain segment in the melt. Under the assumptions put forward, the molar fraction of crystals is simply given by

$$wc = (1 - M/3) \sum_{y(T_{My})}^{\infty} \varphi_y. \tag{34}$$

To formulate the average thickness of the crystallized microphases by

$$\langle yc \rangle = (1 - M/3) \sum_{y(T_{My})}^{\infty} y\,n_y \,/\, \sum n_y$$

$$= (1 - M/3) \sum_{y(T_{My})}^{\infty} y\,x_y \tag{35}$$

is a consequence of relation (30), now explicitly expressing how the thickness distribution of the extended-chain lamellae is determined by the molecular-weight distribution. Within such eutectoid multi-component systems the melting process is characterized by the successive and consecutive melting of fractions of mixed crystals, starting with the lamellae of smallest thickness [8, 10, 11]. We are thus straight forwardly led to the conclusion:

By analysing the melting behavior in eutectoid multi-component systems with extended-chain crystals, we have the outstanding possibility of determining the chain-length distribution within the range of $y_{min} \leqq y \leqq y_{max}$.

The excellent fit of calculations to the experimental results which is shown in Figure 31 must be taken as a full justification of this idea. Dependent upon the molecular conditions, the theory correctly describes phenomena like the depression of the maximum melting temperatures, the typical bending of the $wc(T)$ curves and the final maximum degree of crystallinity. It is predicted that

eutectoid multi-component systems with extended chain crystals which are defect saturated can never be brought into the perfectly crystallized state of matter.

Having a chain-length distribution with not too short molecules, the maximum degree of crystallinity is obtained from Equation (33) to be equal to

$$wc_{max} = 1 - M/3.\tag{36}$$

Hence, the maximum degree of order is predicted to be uniquely determined by the defect parameter M independent of the average thickness of the lamellae considered.

Summarizing we come to the following statements:

— oligomer multi-component systems with extended-chain crystals are eutectoid multi-component systems

— the "primary equilibrium colloid-structure" is therefore uniquely determined by the molecular-weight distribution

— in multi-component systems extended-chain mixed microphases always seem to become defect saturated. The can be considered to represent autonomous non-homogeneous microphases, each of them being thermodynamically equivalent

— crystallization segregation runs within equivalent subvolumina so as to optimize the possibilities of approximating to thermodynamical equilibrium even in multi-component systems.

Non-equilibrium phenomena [4, 8, 15] have been observed. Very small modifications of the composition within many of the coexisting microphases may easily have consequences that can be observed on a macroscopic scale. This is one of the most important reasons behind the reliable "deformations of cp-melting curves" which can easily be produced by an appropriate annealing program. A method to minimize these non-equilibrium phenomena is to continuously cool the sample down so as to enforce comparable kinetic conditions in all stages of crystallization. This proposal is defended by the observation that most of the deformation of the *cp-* curves due to annealing procedures leads to a kind of "oscillation" in respect to the measurement performed after having the sample slowly and continuously cooled down.

Crystallization of networks

The unique structural feature of all permanent networks is the presence of chemical crosslinks. Assuming stereoregularity in the chemical chain structure, it is evident that the crosslinks themselves are indeed "non-crystallizable units" [15–18]. The chains between two crosslinks being considered as crystallizable sequences, we may in principle treat networks with a chain-length distribution as eutectoid multi-compo-

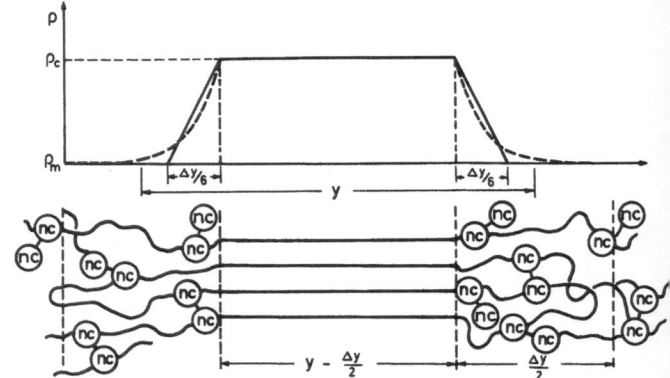

Fig. 32. Model of an extended-sequence mixed crystal within a crystallizable network. This picture (as well as that in Fig. 28) supports the idea of a thermodynamically non-homogeneous and autonomous microphase as manifestly plausible

nent systems. We only need to define the chains of different length as "particles of the multi-component system". Doing this, we have to keep in mind that these particles are not at all autonomous due to their being bound to the network. Chain diffusion is expected to become seriously constrained.

Fortunately, we should expect a tendency to increase the compatibility within the non-homogeneous microphases since we now have crosslinks inserted into the boundaries instead of "energetically very expensive" chain-ends (see Fig. 32). Not knowing the molecular situation, let us empirically modify the melting theory hoping that herewith effects of network-constraints are also accounted for. To define the maximum disparity of chain length by [7, 15–18]

$$\Delta y = M\, y(T) + B(xc)\tag{37}$$

was found to be very adequate whereby

$$B(xc) = B0\, xc/(1 - xc) + B1\tag{38}$$

xc is the molar fraction of the crystallizable units. The parameter A is assumed to be identical with the value found for eutectoid oligomer systems. $B0$ and $B1$ are considered as system-typical constants which also account for "network constraints" including those developed on crystallization.

Having a statistically random chain-length distribution, the molar mass fraction of crystals is equal to

$$wc = Ao\, xc^{(y-1)}\, ((y - yk)\,(1 - xc) + xc)$$

$$Ao = 1 - M/3\tag{39}$$

Table 4. Maximum relative disparity of chain-lengths

$y_k = y$; $B_0 = .9$; $B_1 = 50$; $M = .15$

xc	Δy	yk	$w_{c\,max}$	$\Delta y/\langle yc \rangle$	$\langle yc \rangle_o$
.98	99	33	.49	1.2	82
.96	76	25	.35	1.54	49
.94	67	22	.24	1.79	38
.90	61	20	.12	2.1	29

$\langle yc \rangle_o = x_c/(1 - x_c) + y_k$

with

$$yk = (B(xc)/3 - 1/2)/Ao \qquad (40)$$

characterizing the average number of "non-crystallized" a priori crystallizable units per y-sequence. It is important to recognize that yk is proportional to the internal parameter $B(xc)$.

To increase the value of $B(xc)$ is equivalent as to have growing solubilities in the mixed extended-chain lamellae. Let us illustrate the situation in polyethylene networks by comparing the maximum disparity in chain length with the average "crystal-thickness" defined in terms of a two-phase model [8, 16, 18]

$$\langle yc \rangle = Ao\,(xc/(1 - xc) + y - yk). \qquad (41)$$

In the extreme limits $y = yk$, we find that the maximum disparity of chain-segment lengths exceeds the average crystal thickness $\langle yc \rangle_o$ in each case (Table 4). Crystallization segregation of c-sequences should evidently be possible under very liberal conditions. Due to extremely large widths of the distorted boundaries, at the final steps of crystallization most of the

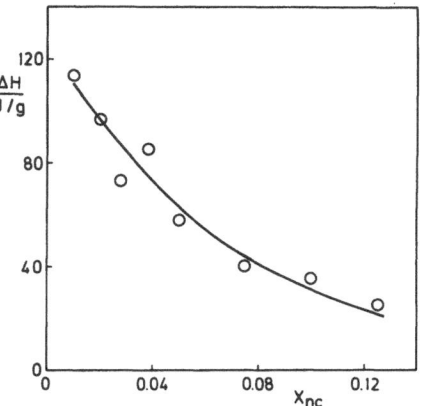

Fig. 33. The melting enthalpy of radiation-crosslinked linear polyethylene against the molar fraction of non-crystallizable units that is uniquely related to the density of crosslinks. The solid line is computed with the parameters compiled in Table 4

c-sequences left in the amorphous layers can "crystallize". Only very poor rearrangements are necessary to realize the required defect segregation. The size of the equivalent volume elements within each of which all of the crystallization rearrangements occur, is therefore extremely small, becoming at least nearly identical to the volume occupied by the finally crystallizing extended-c-sequence lamellae themselves.

To have so much distorted extended-c-sequence lamellae is at the expense of "crystallinity". The molar mass fraction of crystals should therefore become greatly reduced when the density of crosslinks is increased (Fig. 33). For radiation-crosslinked polyethylene this hypothesis is found to hold true, and calculations indeed agree surprisingly well with the experimental results [15–18] (see also Table 5).

Table 5. Experimental and theoretical data of crosslinked PE [59]

$\Delta h(T) = 970$ cal/mole/unit; $T = 418$ K, $\Delta C = 1.4$ cal/mole/unit/degree $2\sigma_e/\Delta h = 2.05$; $M = .15$; $B = .9$; $B1 = 40$

Dosis M_{rad}	x_{nc} mole %	ΔH_{exp} J/g	ΔH J/g	w_c mol %	$\langle y_c \rangle_{exp}$ nm	$\langle y_c \rangle$ nm	G_{exp} MPa	G MPa
500	.125	25	21	.12			6.5	6.7
400	.1	35	31	.17	.9	1.1	6.1	5.9
300	.075	40	44	.24			5.5	5.1
200	.05	58	63	.35	2.5	2.4		
150	.038	85	75	.41			2.7	3.3
110	.028	73	86	.47	4.8	4.4	2.6	2.7
75	.02	96	96	.53				

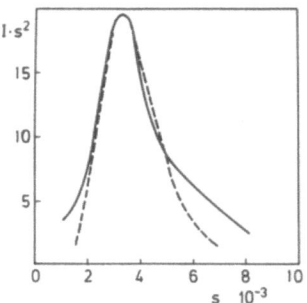

Fig. 34. SAXS pattern of radiation-crosslinked polyethylene (x_{nc} = .025) at 360 K according to [15,16]. The solid line is computed under the assumption that the internal structure within each of the clusters has been fully reorganized on heating so that there are no traces back to the super structure observed at lower temperatures [15,16]

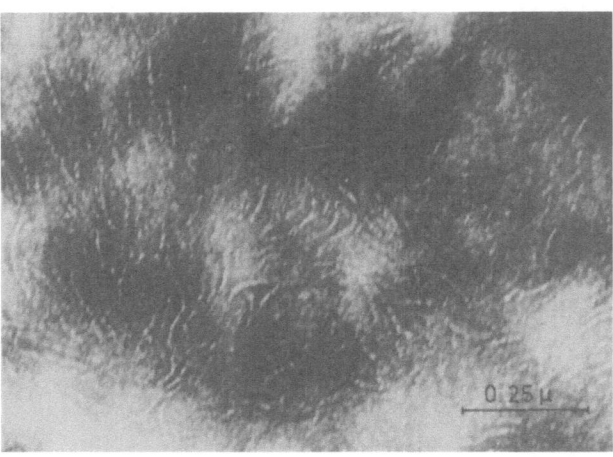

Fig. 35. Electron micrograph of an ultra-thin cut of a 50 Mrad sample stained at 313 K [57,58] according to [15,16]. The lamellae are seen to be comprised of smal blocks

The superstructure in networks only comprising extended-chain-sequence crystals (at not too low degrees of crosslinking), is characterized by a statistically random spatial distribution of the microphases over the total ensemble of relatively small clusters. The SAXS pattern can be calculated under the assumption of having a "cluster-gas" showing no cluster-cluster distance correlations [16]. Moreover, after partial selective melting at elevated temperatures, calculations could be fitted to the experimental data only on the use of a statistically homogeneous cluster gas model. Hence, there are no memory effect saving elements of the superstructure present at lower temperatures (Fig. 34). This can be taken as evidence that all of these processes happen to occur which are required to bring each of the existing microphases nearer to its equilibrium state. These processes are accompanied by a full reorganization of the cluster structure.

To directly confirm characteristics of the colloid structure, an electron-micrograph of a polyethylene network (xc = 0.975) is depicted in Figure 35. The samples have been etched and contrasted according to Kanig's method [57,58]. Each of the coherence clusters is comprised of only about two lamellae, which compare fairly well with the size of the correlation parameter (N = 2) which has been obtained from the description of the SAXS pattern with the aid of Hosemann's linear theory of paracrystalline structures [16].

By the excellent theoretical representation of the melting process for differently crosslinked polyethylene (Fig. 36) we find the assumptions concerning the nature of the crystallization segregation in fact fully defended. Moreover, we are led to the conclusion that

the chain-length distribution within radiation-melt-crosslinked polyethylene networks should strictly correspond to a statistically random distribution. It is worth remarking at this point on the fact that our method represents to our knowledge the only experimental possibility to directly determine the chain-length distribution in networks within the range of $y >$ 15–18.

The physical ideas behind our model are also confirmed by the theoretical representation of additional experimental results. These results are all compiled in Table 5.

Final comments

In the present work we have been concerned with the fundamental aspects in which the crystallization of oligomers and eutectoid copolymers may be considered to be topologically the same. The emphasis was placed on the phenomenon of crystallization segregation of "non-lattice compatible chemical units" like chain-ends, short-chain branchings or crosslinking units. Crystallization and melting of exended-chain multi-component systems can then be formulated in very general terms. This arises from the abstract manner in which the defect situation within mixed extended-chain or extended-chain-segment lamellae is thermodynamically disposed of. Breaking away from some of the classical principles by postulating these non-homogeneous lamellae to represent thermody-

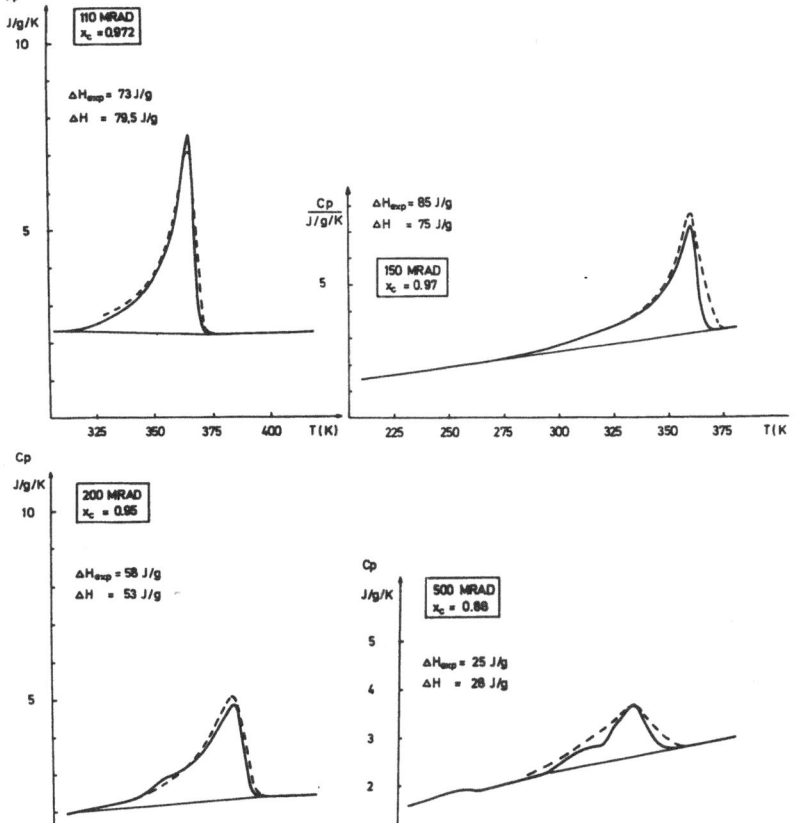

Fig. 36. DSC data of various polyethylenes radiation crosslinked in the melt. The dotted lines have been computed using the parameters listed in Table 4 [59].

namically autonomous and equivalent microphases, a thermodynamics of eutectoid multi-component systems can be developed. We are then led to a proper understanding of melting accompanied with the recognition that the "primary colloid structure" is uniquely related to the chemical conditions within the eutectoid multi-component systems. An attraction of the model developed is that it allows to come to a quantitative representation of various properties within the multi-phase range of such different systems as *n*-alkane multi-component systems and polyethylene networks.

Danksagung

Wir danken der Deutschen Forschungs-Gemeinschaft für die großzügige Förderung.

References

1. Flory PJ (1955) Trans Farad Soc 51:848
2. Roe J, Smith KJ, Krigbaum WR (1961) J Chem Phys 35:1306
3. Krigbaum WR, Roe RJ, Smith KJ (1964) J Polymer 5:533
4. Wunderlich B (1980) Macromolecular Physics Vol 3, Academic Press, New York
5. Mandelkern L (1964) Crystallization of Polymers, McGraw-Hill Book Comp, New York
6. Kilian HG (1965) Koll Z Polym 202:97
7. Glenz W, Kilian HG, Klattenhoff D, Stracke F (1977) J Polymer 18:685
8. Kilian HG (1968) Makromol Chem 116:219
9. Asbach GI, Drexhage KH, Heidemann G, Glenz W, Kilian HG (1970) Makromol Chem 139:115
10. Asbach GI, Kilian HG, Stracke F (1982) Coll Polym Sci 260:151
11. Asbach GI (1972) Thesis, Univers Ulm, Abt Exp Phys
12. Stracke F (1975) Thesis, Unvers Ulm, Abt Exp Phys
13. Neppert BT (1985) Thesis, Univers Ulm, Abt Exp Phys
14. Asbach GI, Geiger K (1979) Coll Polym Sci 257:1049
15. Schmidt H (1982) Thesis, Univers Ulm, Abt Exp Phys

16. Heise B, Kilian HG, Schmidt H (1981) Coll Polym Sci 259:611
17. Holl B, Kilian HG, Yeh GSY (1985) Coll Polym Sci 263:313
18. Kilian HG, Unseld K (1986) Coll Polym Sci (in press)
19. Haase R (1956) Termodynamik der Mischphasen, Springer-Verlag, Berlin
20. Callen HB (1960) Thermodynamics. Wiley Int Ed, New York
21. Buckley CP, Kovacs AJ (1984) Structure of Crystalline Polymers (Hall IH) 261
22. Scherr H, Hägele PC, Großmann HP (1974) Coll & Polym Sci 252:871
23. Zahn H, Pieper ? (1962) Koll Z Z Polym 180:97
24. Kern W, Davidovits K, Rauterkus J, Schmidt GF (1961) Makromol Chem 43:106
25. Flory PJ, Vrij A (1963) J Am Chem Soc 85:3548
26. Broadhurst HG (1962) J Res Nat Bur Standards 66A:241
27. Kitaigorodskij AI (1961) Organic Chemical Crystallography, Consultants Bureau, New York
28. Turner B (1971) Ind Eng Chem Prod Res Develop 10:238
29. Mnyukh Y (1960) Zh Strukt Khim 1:370
30. Rinnaudo C, Aquilino D (1979) Acta Cryst 35A:992
31. Strobl G (1978) Coll Polym Sci 256:427
32. Müller A (1932) Proc Roy Soc 138A:514
33. Strobl G, Ewen ? (1974) J Chem Phys 61:5257
34. Hoffman JD (1952) J Chem Phys 20:541
35. Ohlberg SM (1959) J Phys Chem 63:248
36. Hoffman JA, Smyth ChP (1950) J Am Chem Soc 72:171
37. Finke HL, Gross ME, Waddington G, Hoffman HM (1953) J Am Chem Soc 76:333
38. Flory PJ (1953) Principles of Polymer Chemistry. Cornell University Press, Ithaca, New York
39. Nechitailo NA (1960) Zh Fiz Khim 34:2694
40. Gilg B, Skoulios A (1971) Makromol Chem 140:149
41. Smith P, Manley R St John (1979) Macromolecules 12:483
42. Flory PJ (1941) J Chem Phys 9:660
43. Flory PJ (1942) J Chem Phys 10:51
44. Huggins ML (1941) J Chem Phys 9:440
45. Huggins ML (1942) Ann NY Acad Sci 43:1
46. Stavermann AJ, van Sauten JH (1941) Rec Trav Chim 60:76
47. Stavermann AJ (1941) Rec Trav Chim 60:640
48. Oleinik EF, Finishing Lapping Parte P 1800188c
49. van der Waals JH, Hermanns JJ (1950) Rec Trav Chim 69:949
50. Stuart HA (1953) Die Physik der Hochpolymeren, Bd II, Springer-Verlag, Berlin
51. Prime RB, Wunderlich B (1969) J Polym Sci Phys Ed 7:2073
52. Prime RB, Wunderlich B, Melillo L (1969) J Polym Sci Phys Ed 7:2091
53. Metha A, Wunderlich B (1975) Coll Polym Sci 253:193
54. Anderson FR (1965) J Polym Sci, Part C 8:275
55. Hosemann R, Bagchi SN (1962) Direct Analysis of Diffraction by Matter. North Holland Publ Comp, Amsterdam
56. Schmidt H, unpublished calculations
57. Kanig G (1973) Koll Z Z Polym 251:782
58. Kanig G (1974) Kunststoffe 64:474
59. Kilian H-G, Unseld K, Jaeger E, Müller J, Jungnickel B (1985) Coll & Polym Sci 263:607

Received February 15, 1986;
accepted February 28, 1986

Author's address:

Prof. Dr. H.-G. Kilian
Abteilung für Experimentelle Physik
Universität Ulm
Oberer Eselsberg
D-7900 Ulm, F.R.G.

Progress in Colloid & Polymer Science   Progr Colloid & Polymer Sci 72:83–96 (1986)

Structure and properties of polyamide 12 alloys*)

G. Goldbach, M. Kita[1]), K. Meyer, and K. P. Richter

Research and Development Centre of Hüls Aktiengesellschaft, Marl, F.R.G., and
[1]) Plastics Research Laboratory of Daicel Chemical Industries, Ltd., Shinzaike Aboshi-Ku, Japan

Abstract: Polymer alloys composed of polylaurolactam (polyamide 12) and segmented polylaurolactam with different content of soft segments are interesting engineering thermoplastic materials having a broad spectrum of properties. The reason for this broad spectrum are the phase structure and the special interactions between the components at the phase boundaries: From scanning electron microscopy and torsion vibration analysis it is to be concluded, that the components are thermodynamically incompatible with one another. High resolution electron microscopy, however, shows that there is a strong interaction between the components at the phase boundaries. The bond between the components is formed by crystalline lamellae which seem to originate from both components: from the hard segments of the segmented polylaurolactam and from the polyamide 12 molecules.

Volume/strain and stress/strain experiments indicate that the strength of the phase boundaries depends on the composition of the alloys. At low and high contents of one component the boundary strength is high and failure of these alloys takes place by shear, leading to high elongations at break. In the range of middle compositions microcrack formation is observed, causing a considerable reduction of the elongation at break.

Key words: Polyamide, polymer alloy, phase structure, deformation mechanism, volume/strain properties.

1. General

The intensive research and development widely observed today in the area of high molecular weight materials are comparable in terms of scope and objectives with the research and development observed in metallurgy since about the turn of the century. What is meant here is the development of new materials by alloying. In metallurgy, alloying is the mixing or bonding of two or more types of metal atoms to form systems with new properties. The individual components of the alloy may be infinitely soluble in one another (solid solution) or may exhibit a greater or lesser degree of separation [1]. The term polymeric alloy is defined in exactly the same way [2].

There are two noteworthy reasons for the intensive work being carried out today in the field of polymeric alloys. On the one hand, by combining known and proven polymeric materials it is possible to produce new property profiles. To the forefront in many cases is the improvement in the mechanical properties, such as impact strength, deflection temperature under load, and resilience, although it is also possible to obtain defined electrical and optical characteristics by alloying. On the other hand, the time required for developing polymeric alloys, which is about 3–5 years, is substantially less than the time involved in, for example, the development of a new polymeric material.

2. Alloy components

An important precondition for good alloyability is a certain interaction between the components to be alloyed. This applies in particular to polymeric alloys

*) Plenary lecture at the 32nd general meeting of the Kolloid-Gesellschaft e. V. and the "Berliner Polymeren-Tage" 2–4 October 1985.

since, for entropic reasons, the various macromolecular components are generally present as separate phases. Interactions between the alloy components are therefore only possible at the phase boundaries. Hence, boundary interactions are a critical factor in the choice of components.

In the present work, components chosen have a chemical structure which suggests that there will be a certain interaction between them. One component is polylaurolactam

$$R-(N-(CH_2)_{11}-C)_n-R' \quad (R, R' \text{ end-groups})$$
$$\quad\ \ |\qquad\qquad\ \ ||$$
$$\quad\ \ H\qquad\qquad\ O$$

a high molecular weight material which, under the name polyamide 12 (PA 12), has interesting applications particularly in industry [3–5].

The other component in each belongs to a series of so-called segmented polymers which have the following chemical structure:

One building block consists of oligolaurolactam (i.e. $l=10$). It can be crystallized, and is referred to as the hard segment because of its high melting point.

The other building block consists of oligotetrahydrofuran ($k=4$, number average of molecular weight 1000 g/mole). Because of its low glass transition temperature and low degree of crystallinity, it constitutes the so-called soft segments. The hard and soft segments are linked via an ester bond brought about by dicarboxylic acid.

Such systems are already state of the art today and are considered as thermoplastic polyamide elastomers with widely variable properties [6, 7].

The basic concept behind the choice of these alloy components is obvious: the fact that the laurolactam sequences of the segment polymers are chemically "identical" to the molecules of polyamide 12, so that the components can be expected to exhibit a certain degree of interaction.

Characteristic structural parameters of the components mentioned are summarized in Table 1.

Table 1. Structural parameters of the components

$[\eta]^{a)}$, ml/g of the component	Weight fraction of hard segments	$\bar{M}n$ (g/mol) of hard segments	$\bar{M}n$ (g/mol) of soft segments
—	100	—	—
122	90	11 000	1 000
140	85	6 800	1 000
134	75	4 100	1 000
157	60	2 400	1 000
158	50	2 000	1 000
137	40	1 500	1 000

$^{a)}$ determined at 25 °C; solvent: Hexafluoro-iso-propanol

Against this background, the following questions will be discussed:
— What sort of superstructure do such alloys have?
— What are the structures of the phase boundaries?
— How are the properties, particularly the mechanical properties, affected by the superstructure and phase adhesion?

Before investigating these problems in more detail, it is expedient first to consider the morphology and some properties of the starting components themselves. The experimental techniques used here and the preparation of the materials are described in [6, 7].

3. Properties of the alloy components

3.1 Crystalline phases

The properties which will initially be discussed are those which provide information about the crystalline phases. To this end, Figures 1a and 1b each show two graphs, a torsional vibration diagram (in the form of the loss modulus) and a DSC fusion thermogram over the common temperature axis. The torsional vibration diagram shows the glass transition temperature as the important characteristic feature of the amorphous phase, while the fusion thermogram shows the melting range with the melting peak typical of the crystalline phase. In the case of the oligotetrahydrofuran, the glass transition temperature is about −70 °C and the melting point 24 °C; for polyamide 12, the corresponding temperatures are 45 °C and about 178 °C (anhydrous samples).

Figure 2 shows the DSC fusion thermograms for a series of polyether ester amides (PEEA), the oligotetrahydrofuran content increasing from the upper curve to the lower curve. The top thermogram shows

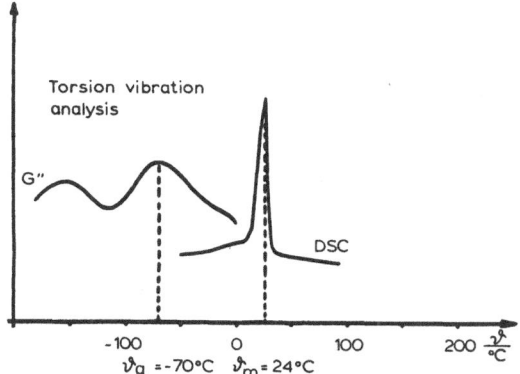

Fig. 1a. Loss modulus/temperature diagram and DSC melting thermogram of oligotetrahydrofuran

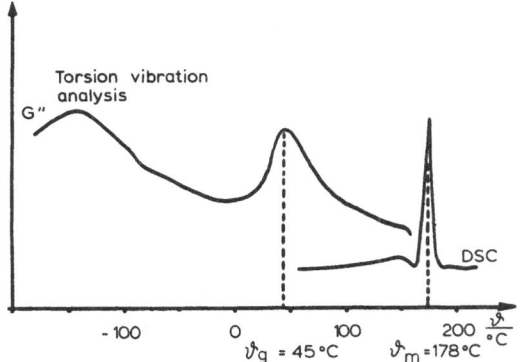

Fig. 1b. Loss modululs/temperature diagram and DSC melting thermogram of PA 12

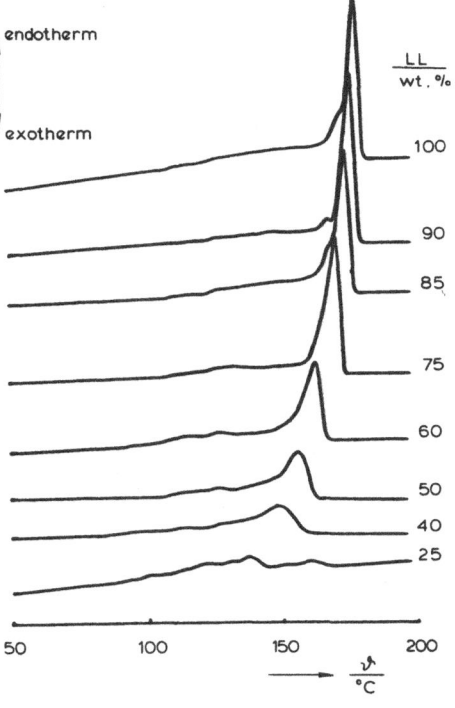

Fig. 2. DSC melting thermograms of PA 12 and a series of PEEA of different content of oligolaurolactam structural units

the fusion of pure polyamide 12. The fact that it fits easily into the series indicates that the process involved in the segmented polymers is the melting of crystalline oligolaurolactam regions.

This is also confirmed by, for example, wide angle X-ray scattering.

Two systematic effects can be observed:

1. a decrease in the heat of fusion (represented by the area under each peak) and

2. a fall in the melting point (represented by the position of the maximum).

Since the process involved is the melting of oligolaurolactam crystallites, it is reasonable to relate the heat of fusion to the laurolactam content. It is found that all materials have virtually the same value, which is only slightly lower than that of pure polyamide 12. This means that the decrease in the heat of fusion is essentially a dilution effect.

Figure 3 shows this result in the form of a graph.

The following additional comments may be made about the decrease in the melting point: the mean chain length of the soft segments is 1000 g/mole and is identical for all samples, as already stated. If the amount of soft segments is increased, the result is that the mean

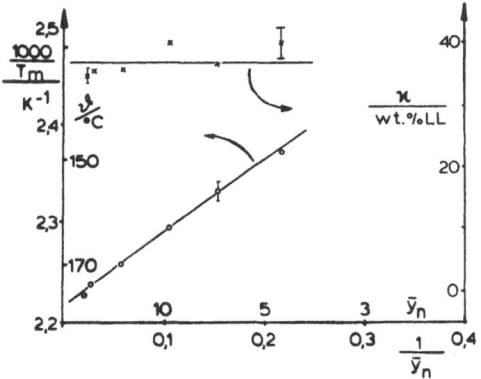

Fig. 3. Dependence of crystallinity x calculated from the heat of melting and the reciprocal absolute melting temperatuure (peak maximum) on the average degree of condensation of oligoamid structural units $\bar{y}_n$ ($\times$: crystallinity, O: $\bar{y}_n$)

chain length of the hard segments is reduced. Accordingly, the series shown in Figure 3 comprises systems with a constantly decreasing mean chain length of the hard segments.

If, in accordance with Flory [8], the reciprocal absolute melting point is plotted against the number average chain length, a straight line is obtained, provided that the melting point depends only on the chain length.

Figure 3 shows that this is the case. In other words, the building blocks of the hard phase melt as though they were present as oligomers and were not linked by soft segments (for a more detailed discussion of the melting behaviour, see [7]).

These results can be used to produce a certain picture of the crystalline structure of the polyether ester amides. The polyether ester amides contain a crystalline polyamide phase comprising about 25–30 % of the oligoamide present. This means that the oligolaurolactam sequences in the polyether ester amides show obviously the same extent of crystallization as the pure homopolymer. The melting point of these crystalline phases is related to the molecular weight according to the Flory theory.

It may be added that the oligotetrahydrofuran, which in the form of the pure component crystallizes relatively easily, exhibits only very small crystallinity when the molecules are incorporated as segments into the chain. Instead, it forms part of the amorphous regions, which will be discussed below.

3.2 Amorphous phases

Figure 4 shows the temperature dependence of the loss modulus (G'') for a series of polyether ester amides, as well as the curves for the two pure components polyamide 12 and oligotetrahydrofuran. The results were obtained using anhydrous samples.

The upper curve applies to pure polyamide 12. As already stated, the maximum at 45 °C is attributable to the glass transition temperature of this material, i. e. the amorphous regions exhibit a transition at 45 °C and the measurement frequency of 1 Hz used here. The lower curve was obtained with a pure oligotetrahydrofuran which has about the same molecular weight as the soft segments in the polyether ester amides discussed here. Its glass transition temperature is −75 °C. If the curves are compared, it will be seen that the glass transition temperature of the polyamide 12 is shifted towards lower temperatures by incorporation of soft segments into the macromolecule, and this effect is

accompanied by the appearance of a maximum between −60 °C and −68 °C, depending on the overall composition. If the amount of soft segments is not too high, two maxima occur. We conclude from this that in this case and in the other samples too there are at least two amorphous phases of different compositions. The phase which has the lower glass transition temperature is *rich* in oligotetrahydrofuran, whereas the other phase which has a higher glass transition temperature is rich in oligolaurolactam. So the oligotetrahydrofuran not only occurs as a constituent of the laurolactam-rich phase, depressing the glass transition temperature of the latter, but also forms a phase of its own; it can therefore be considered as a partially separated internal plasticizer (the separated part having a low glass transition temperature and therefore acting as a toughener; cf. [7]).

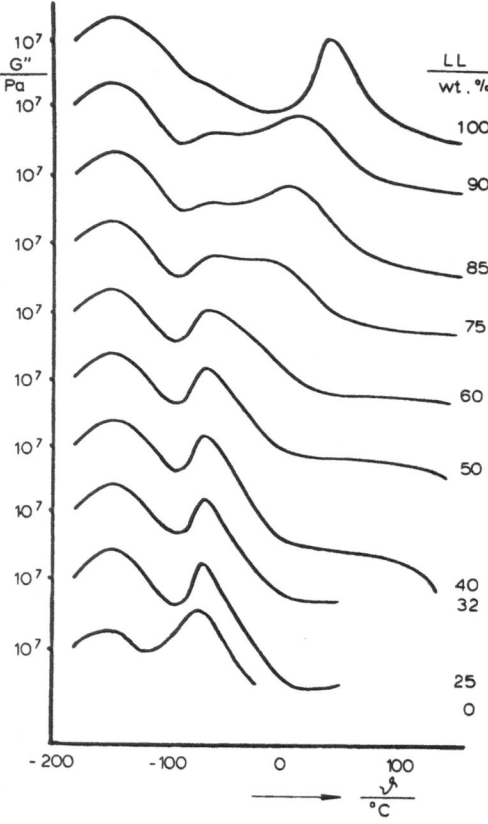

Fig. 4. Loss modulus G'' of PA 12, oligotetrahydrofuran and a series of PEEA of different content of oligolaurolactam structural units as a function of temperature (torsional vibration analysis, frequency 1c/s)

3.3 Superstructure

The various alloy components form superstructures which can be readily observed, particularly under the electronmicroscope.

Polyamide 12 forms dendritic or spherulitic crystals and has a crystallinity of about 30 %. Figure 5 a shows an electron micrograph of a polyamide 12 spherulite as formed from the melt under the usual cooling conditions. Figure 5 b shows that the spherulite has a lamellar structure. The white stripes are the crystalline lamellae, and the black stripes are the "amorphous" regions formed from non-crystalline polylaurolactam. Contrast was achieved as described in [9], by treating very thin films with allylamine/OsO$_4$. The thickness of the lamellae is 6–8 nm and, surprisingly, is completely independent of the thermal history. For example, the fact that some of the samples have been cooled extremely rapidly from the melt or annealed at high temperatures, for example in the region of the crystallization temperature, is unimportant; the lamellae are always found to be 6–8 nm thick. Assuming that the chains are oriented at right angles to the surfaces of the lamellae, 4 basic molecules are incorporated in the lamellae before the chain either folds back on itself or enters the amorphous phase. Lamellae whose thickness corresponds to less than 3 basic molecules are virtually completely absent.

A point of interest is the fact that the hard segments of the polyether ester amides crystallize in the same way as pure polymaide 12 itself. They form spherulitic and dendritic superstructures, more "dilute" structures being obtained as the content of soft segments increases (Figs. 6a–c).

Figure 7 shows the lamellar structure of a segmented polyamide containing 50 % by weight of hard segments. In this case too, the thickness of the lamellae is 6–8 nm, regardless of the thermal history. The dark stripes between the (pale) lamellae are the amorphous "boundary layers" of the lamellae. It is assumed that they consist of non-crystalline oligolaurolactam and oligotetrahydrofuran segments. The larger pale areas are also amorphous zones. They form the coherent phase.

In products with an even higher content of soft segments the superstructure may be lost to such an extent that the lamellae are randomly distributed. Figure 13 shows a photograph depicting the situation. It is explained in detail in section 4.3.1.

As described in the discussion of the torsional vibration curves, the polyether ester amides can possess at

Fig. 5 a. Transmission electron micrograph of a thin melt crystallized spherulite of PA 12 (thin melt film, contrast enhancement by *Pt/C* shadowing)

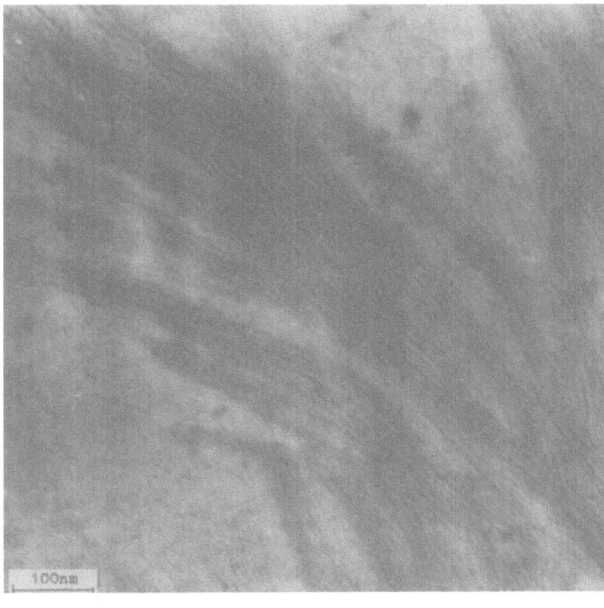

Fig. 5 b. Transmission electron micrograph of a thin melt crystallized film of PA 12; white lines: crystalline lamellae of about 7 nm thickness (contrast by staining with allyl amine/OsO$_4$)

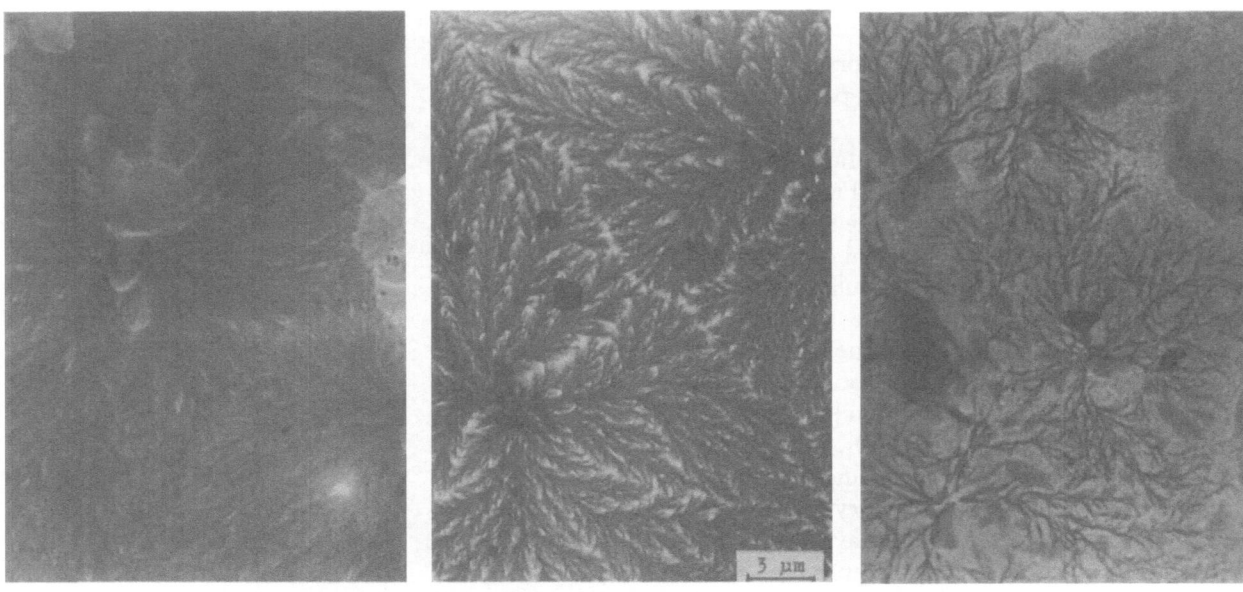

Fig. 6a–c. Transmission electron micrograph of melt crystallized spherulites of PEEA of different content of oligolaurolactam;
a): 75% oligolaurolactam; b): 50% oligolaurolactam; c): 40% oligolaurolactam

least two amorphous phases of different compositions: a laurolactam-rich phase with a higher glass transition temperature and a phase which has a low laurolactam content and whose glass transition temperature is between −60 °C and −68 °C, depending on the composition.

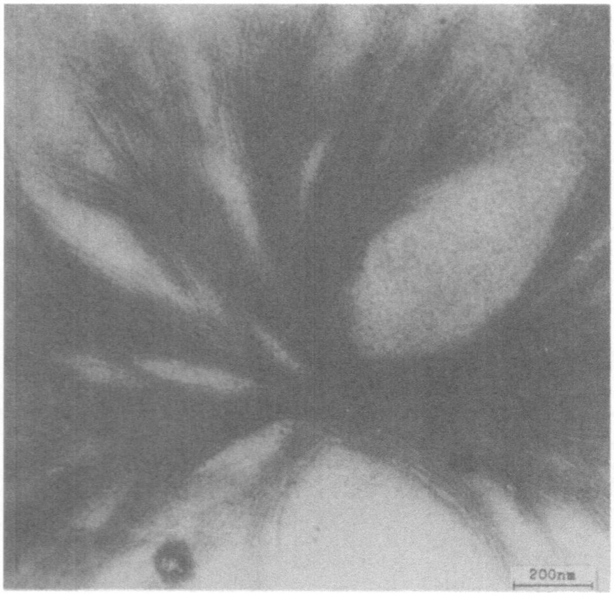

The question naturally arises here as to where these different phases are located and what their morphologies are.

The relevant information is provided by electron microscopic observations [10] (Figs. 8a and 8b).

Figure 8a shows a photograph of a very thin melt film of a sample containing 60% by weight of oligolaurolactam (contrast enhancement by Pt/C shadowing). An interesting morphological phenomenon is observed: embedded in the dendritic superstructure are spherical inclusions of about 0.1–1 μm diameter. We assume that these inclusions are mixed phases which are rich in soft segment, i. e. those phases which give a glass transition temperature at low temperatures in the torsional vibration analysis. It is to be assumed that these phases are formed from oligotetrahydrofuran segments which are bonded to one another principally by shorter non-crystallizable oligolaurolactam sequences. In other words, they are composed of oligotetrahydrofuran molecules or molecular moieties consisting predominantly of oligotetrahydrofuran. The composition can be estimated from the position of the glass transition temperature. The spherical particles consist of about 90% by weight of soft segments and 10% by weight of hard segments.

Fig. 7. Transmission electron micrograph of a thin melt crystallized film of PEEA with 50 wt % oligolauryllactam (contrasting method as in Fig. 5b)

The other amorphous phase which is visible in the torsional vibration diagram has a higher glass transition temperature and is accordingly rich in oligolaurolactam. As the excess phase, it forms the quasi-continuous matrix in which the individual lamellae or dendritic superstructures, including the spherical PTHF-rich particles, are embedded.

There is obviously good adhesion between the matrix and the PTHF-rich particles. This can be seen from Figure 8b, where the structure of the phase boundaries is somewhat clearer. It can be seen that crystalline lamellae from the matrix project into the surface of the particles and thus form a bond between the two. Presumably, hard segments become concentrated at the surface of the oligotetrahydrofuran-rich particles, crystallize when the melt is cooled and then form the stated "lamellar bond" with the matrix.

From the point of view of applications technology, such a morphology is of course advantageous. On the one hand, the soft segments form an amorphous mixed phase with the oligolaurolactam and depress the glass transition temperature. This is the reason for the good elastomeric properties which polyether ester amides possess even at low temperatures. Since the soft segments form part of the chains, there is also no danger of exudation of the "plasticizer".

On the other hand, the soft segments also have an advantageous effect on the mechanical properties in that the segments are rubber-like inclusions and act as a toughener at low temperatures.

4. Alloys

4.1 Preparation

The alloys are prepared by mixing the components, i.e. the pure polyamide 12 and the polyether ester amide, in the molten state in a twin-screw kneader. Mixing was carried out at about 210 °C, this temperature, like the other process parameters, being constant for all systems discussed below.

4.2 Compatibility and component distribution

The question as to whether, and to what extent, the various components are thermodynamically compatible with one another will be discussed first. The inves-

Fig. 8a. Transmission electron micrograph of a melt crystallized film of PEEA with 60 wt% of oligolaurolactam (contrast enhancement by *Pt/C* shadowing)

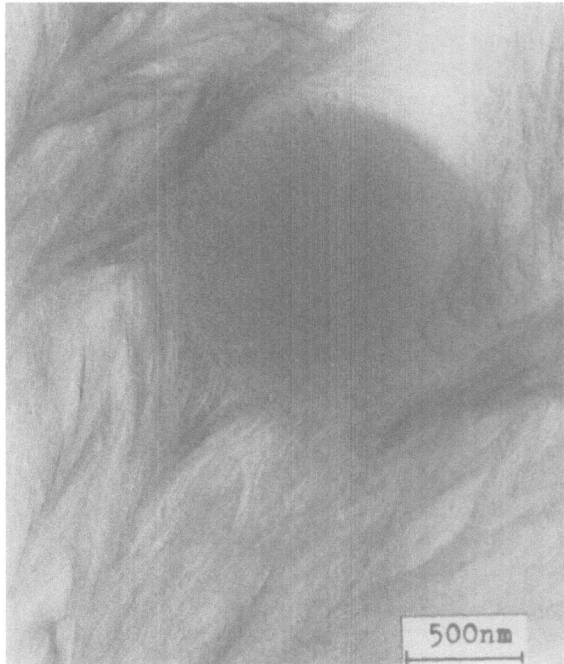

Fig. 8b. Sectional enlargement of Figure 8a

tigations were concerned exclusively with the solid state.

For this purpose, Figure 9 shows a graph in which the loss modulus is plotted as a function of the temperature for alloys which consist of polyamide 12 and a polyether ester amide containing 50 % by weight of soft segments. Each curve shows two clearly separated glass transition temperatures which correspond to the temperatures at which the starting components too undergo a glass transition. This means that the components form separate phases.

This is confirmed by investigations under the scanning electron microscope. For example, Figure 10 shows the distribution of components in an alloy which contains 75 % by weight of polyamide 12 and 25 % by weight of a polyether ester amide containing 50 % by weight of soft segments. The holes in this structure represent the polyamide elastomer, which was dissolved out of the sample by ethanol before the photograph was taken. In this example, the polyamide 12 component forms the continuous phase.

Finally, Figure 11 shows the distribution of the same alloy components as mentioned above, but in this case for a composition consisting of 50 % by weight of polyamide 12 and 50 % by weight of PEEA. Here too, the components are found to be incompatible, the differ-

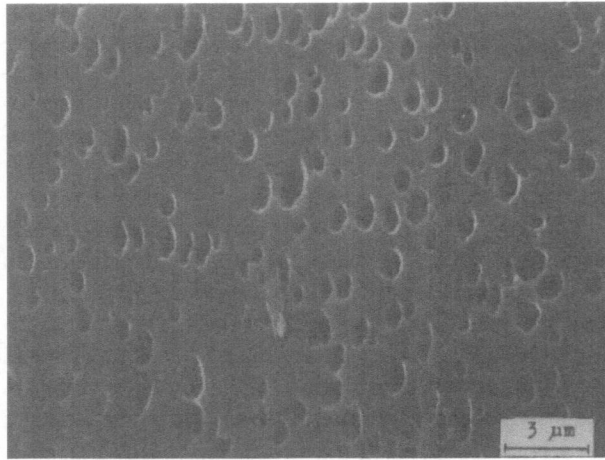

Fig. 10. Scanning electron micrograph of PA 12 based alloy; composition: 75 wt% PA 12, 25 wt% PEEA of 50 wt% oligolaurolactam; the "holes" represent the elastomeric component, which is dissolved by ethanol for contrasting

ence being that both components form virtually continuous phases.

4.3 Phase boundaries

Although the scanning electron micrographs give a detailed picture of the distribution of the components,

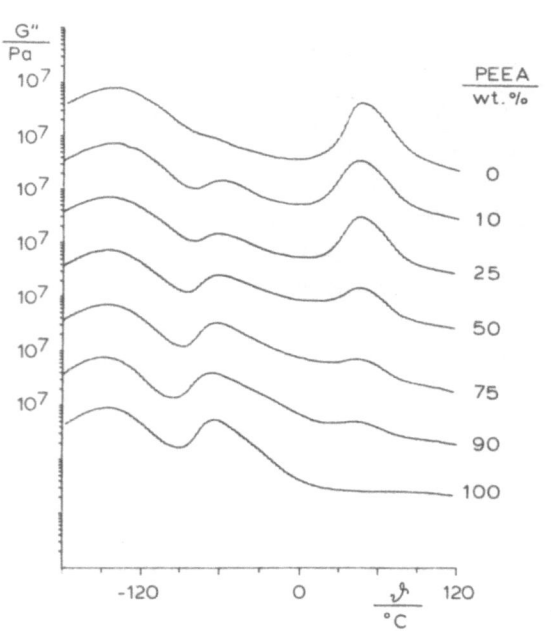

Fig. 9. Loss modulus G'' of PA 12 based alloys; parameter: amount of PEEA of 50 wt % oligolaurolactam

Fig. 11. As shown in Figure 10, however 50 wt% PA 12; both components build nearly continuous phases

they provide little or no information about the phase boundaries themselves, for example about the structure of the phase boundaries and their strength.

This question is, as mentioned above, of particular interest in the development of polymeric alloys, because the applicational properties too depend substantially on the strength of the phase boundaries.

Suitable methods for investigating the phase boundaries of the polyamide 12 alloys are transmission electron microscopy (TEM) and extension dilatometry. High resolution TEM on very thin, "chemically" contrasted films provides a good picture of the structure of the boundaries, while extension dilatometry gives indirect information about their strength.

4.3.1 Boundary structure

Figures 12 a to 12 c show the superstructure of an alloy consisting of 75 % by weight of polyether ester amide (soft segment content 50 % by weight) and 25 % of polyamide 12[1]). The black, spherical regions constitute the polyamide 12, which is the component present in smaller amount and is dispersed in the elastomer phase. The rough boundary between polyamide 12 and the elastomer is itself an indication of intimate dovetailing between the two components (Fig. 12a). This becomes particularly clear under fairly high magnification, and it appears that the bond between the components is formed by crystalline lamellae (Figs. 12c and 13).

The same impression is given even when the polyamide 12 is present as the excess component. In such alloys, the polyamide 12 forms dendritic and spherulitic superstructures which are also linked to one another via the elastomeric polyether ester amide (Fig. 14).

The possible extent of intimate dovetailing between the components is shown in Figure 15. The figures show that on the one hand, individual lamellae project into the elastomer phase over long distances and thus effect dovetailing, while on the other hand the dendrites too link up with one another and thus provide the material with a certain degree of rigidity. The "dovetailed lamellae" are presumably composed of molecular moieties which originate from both components:

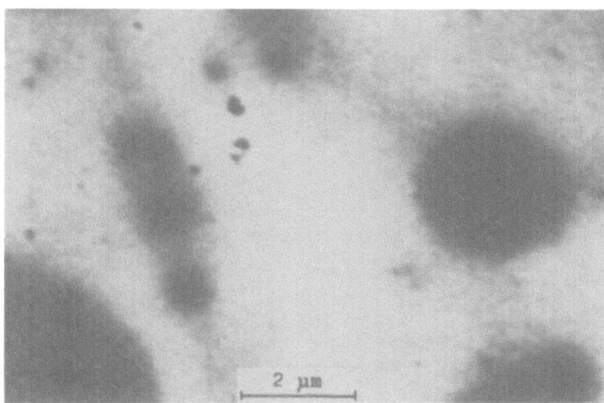

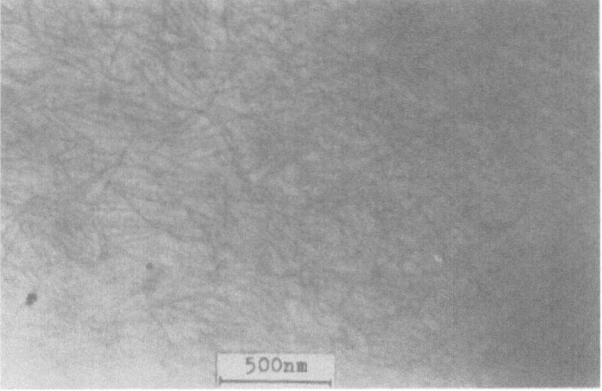

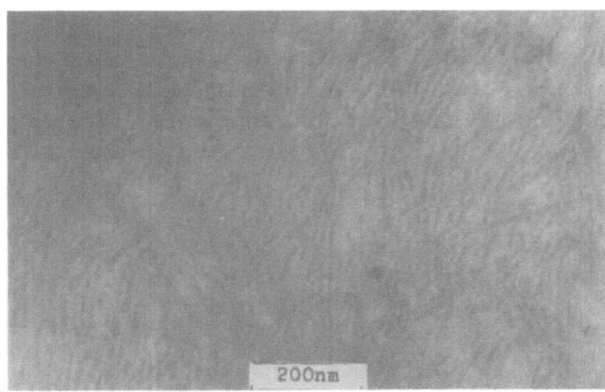

Fig. 12a–c. Transmission electron micrograph of a PA 12 based alloy; composition: 25 wt% PA 12, 75 wt% PEEA of 50 wt% oligolaurolactam; dark areas: dispersed PA 12 phase (contrast by staining with phosphotungstic acid); b) and c): sectional enlargements

from the hard segments of the polyether ester amide and from the polyamide 12 molecules[2]).

---

[1]) Contrast is produced with phosphotungstic acid. Systematic investigations have shown that with this contrasting agent it is possible in the main to bring out structural details (lamellae, amorphous regions) of the polyester amides, while the polyamide component is substantially coloured uniformly grey to black.

---

[2]) This assumption is supported by the experimental finding that the oligolaurolactam sequences of the polyether ester amide component have a higher crystallinity in the alloys than in the pure polyether ester amides.

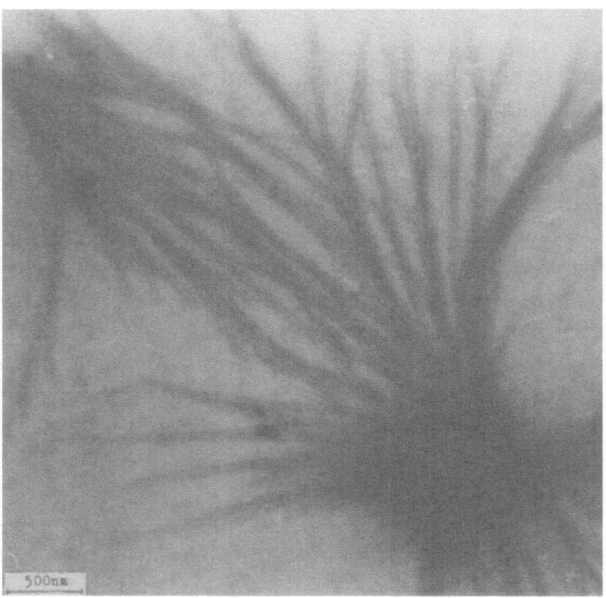

Fig. 13. As shown in Figure 12a, however 50 wt% PA 12, 50 wt% PEEA with 40 wt% oligolaurolactam; the randomly distributed crystalline lamellae seem to be intensively linked to the PA 12 phase

Fig. 15. As shown in Figure 12, however 75 wt% PA 12, 25 wt% PEEA with 50 wt% oligolaurolactam

4.3.2 Boundary strength

The direct determination of boundary strength in polymeric alloys continues to present a problem. Although methods such as interfacial tension measurements or investigations into the so-called tack can provide important additional information in this respect, they are essentially restricted to the study of the separate alloy components. On the other hand, extension dilatometry, first used by Heikens and coworkers [11, 12] to investigate polymeric alloys, gives more extensive information. Although it is only an indirect method, it has the advantage that it can be used for the direct study of "ready-prepared alloys".

Before describing the results in more detail, a few comments will be made about the theoretical background and experimental technique:

The mechanical failure of high molecular weight materials is governed by two failure mechanisms which can be clearly distinguished from one another. One mechanism is flow under shear, also referred to as plastic flow, and the other mechanism is the formation of microcracks or crazes. Both mechanisms have characteristic volume effects: flow under shear takes place at constant volume, whereas micro-crack formation/crazing is associated with an increase in the volume of the sample [13, 14].

Determining the volume during a uniaxial tensile test (extension dilatometry) thus constitutes a suitable experimental technique for obtaining information about failure mechanisms in polymeric alloys [11].

The basic information which can be obtained by means of extension dilatometry is shown in Figure 16. The figure shows (top) *idealized* stress-strain curves for two polymeric materials in which failure is due to

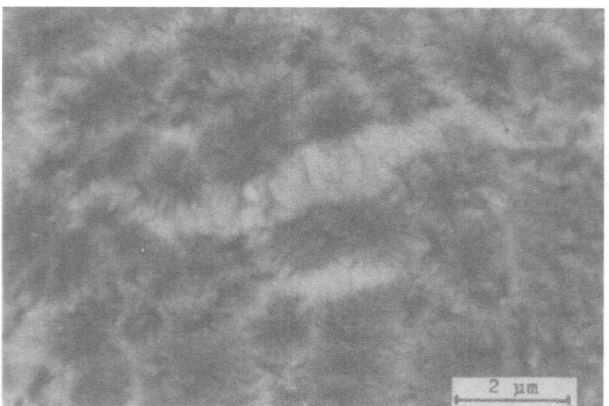

Fig. 14. As shown in Figure 12, however 75 wt% PA 12, 25 wt% PEEA with 75 wt% oligolaurolactam; spherulites are built by PA 12 and linked by crystalline lamellae of PEEA

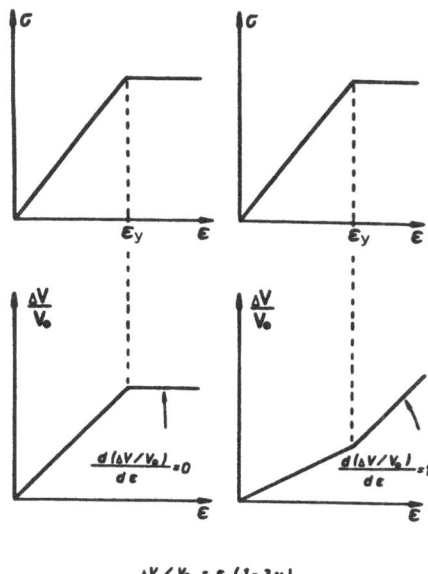

$$\Delta V / V_0 = \varepsilon (1-2\nu)$$

Fig. 16. Schematical stress/ and volume/strain curves of materials of different failure mechanism

flow under shear (example on the left) or microcrack formation/crazing (example on the right). The onset of failure is indicated in each case by the discontinuity in the stress-strain curve, i. e. the so-called yield point.

The lower part of the figure shows the associated volume/strain curves. It can be seen that in both cases the volume increases linearly until the yield point is reached. The slope in these regions is determined by Poisson's ratio ν, in accordance with the stated equation.

When the failure point is reached, the volume/strain curves show characteristic differences. In the example on the left, where failure is due to flow under shear, the curve bends and becomes horizontal, i. e. further extension of the sample up to the point of fracture takes place at constant volume.

On the other hand, in the other example, where failure is initiated by microcrack formation or crazes, the curve bends upwards, i. e. the volume continues to increase with increasing extension. The slope of the curve is therefore a measure of microcrack formation: the greater the number of cracks formed, and the larger these are, the steeper is the slope [11–14].

As mentioned above, the boundary strength plays an important role in determining the quality of an alloy. If the boundary strength is low, separation may

occur at the phase boundaries under mechanical load. Such failure processes result in the formation of cavities or microcracks, which make an additional contribution to the volume.

Assuming that, when considered alone, the alloy components fail exclusively by flow under shear — this applies to the polyamide 12 alloys studied here, as will be shown later — the slope is:

$$S = \frac{d \, \Delta V / V_0}{d \, \varepsilon}.$$

The experimental arrangement for studying the volume/strain behaviour was similar to Heikens' extension dilatometer [11], except that this was modified to permit changes in sample volume in the parts per thousand range to be recorded. The displacement fluid used was water; the sample dimensions of the tensile tests were about $100 \times 10 \times 1 \, mm^3$, and the traversing speed of the tensile test machine was 11 mm/min.

Results of the extension dilatometry studies are shown in Figures 17 to 19.

Firstly, Figure 17 shows the behaviour of the pure components. The upper part of this figure shows the stress-strain curves. The values for elongation at break (not given here) are greater than 300%. Polyamide 12 itself, as well as samples containing up to about 85% by weight of hard segments, show a weak maximum in the stress-strain curve. As the content of soft segments increases, the curves become flatter, particularly in the initial range, signifying the increasing elastomeric character.

The lower part of Figure 17 shows the associated volume/strain curves. For small extensions, the volume initially increases sharply with extension, after which the curves flatten out until finally, at large extensions, virtually no further increase in volume is observed. The Poisson's ratios ν recorded on the curves have been calculated from the linear slope at small extensions. As expected, ν increases as the content of soft segments increases, values of 0.5 being reached for samples containing more than 70% by weight of soft segments. This is an indication of the pronounced elastomeric character of these products.

In spite of the pronounced elastomeric character, the materials still have a sort of yield point, as can be recognized from the flattening of the stress-strain curves. Above these yield points, the associated cubic expansion curves run virtually parallel to the abscissa. The values of the slopes S are noted on the curves.

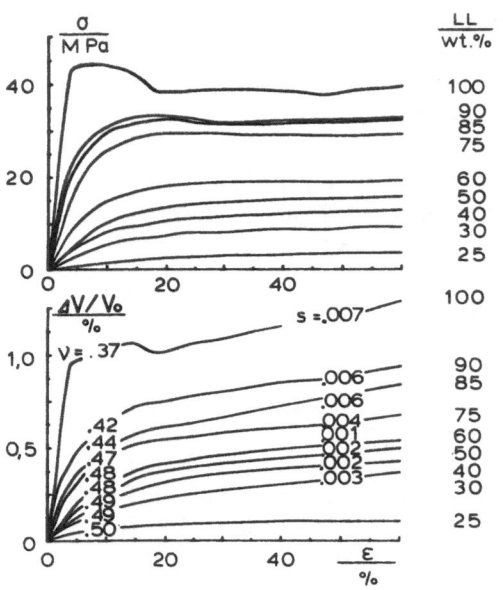

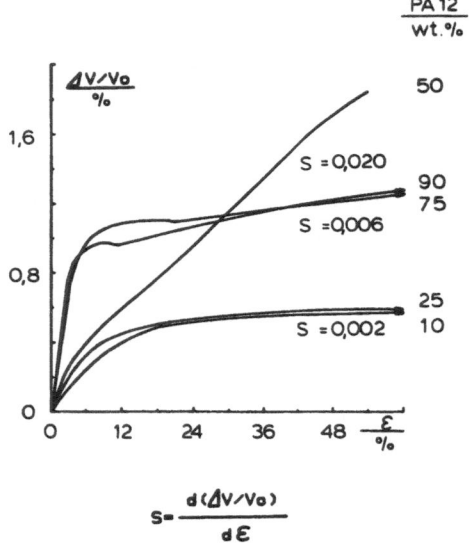

Fig. 19. Volume/strain curves of the PA 12 alloys of Figure 18

Fig. 17. Stress/strain (above) and volume/strain (below) curves of PA 12 and a series of PEEA

The values for S, which are 0.001 and 0.007, are extremely low. This means that failure of the pure alloy components occurs virtually exclusively through shear flow processes. Microcrack formation and crazing are negligible.

Figure 18 shows, by way of example, the stress-strain curves for a series of alloys of polyamide 12 and a polyether ester amide containing 50 % by weight of soft segments. The composition of the alloys varies between 10 % by weight and 90 % by weight of polyamide 12.

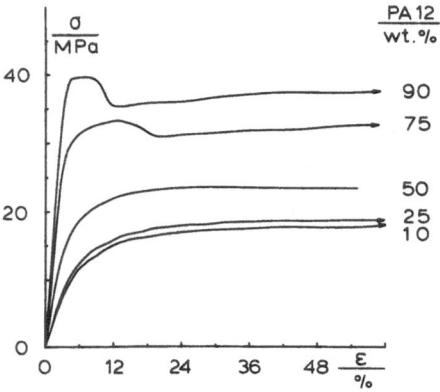

Fig. 18. Stress/strain curves of a series of PA 12 based alloys with a PEEA-component of 50 wt % oligolaurolactam

Samples containing more than 75 % by weight of polyamide 12 still show a pronounced yield point (maximum in the stress-strain curve), which is associated with necking of the sample (Note: in this sample, the polyamide 12 forms the continuous matrix; in alloys having a low polyamide 12 content, on the other hand, the elastomer is the continuous phase (cf. Figs. 10 and 14)).

The associated volume/strain curves are shown in Figure 19. With regard to failure, remarkable behaviour is observed: the curves for alloys which have high and low contents of one component are qualitatively similar to the curves for the pure components. This means that a linear elastic increase in the volume with elongation (the Poisson's ratio ν calculated from this is shown in Figure 20 as a function of the composition) is first observed; this behaviour then changes gradually to a deformation process at constant volume when the yield point is reached.

Consequently, failure of these alloys takes place by shear deformation, as described for the pure components. Surprisingly, this no longer applies to alloys of middle compositions. These exhibit anomalous behaviour. As shown in Figure 19, the volume of a sample containing roughly equal amounts of both components increases virtually linearly above the yield point, until fracture occurs. In other words, in the sample of middle composition, microcracks are formed continuously during plastic deformation.

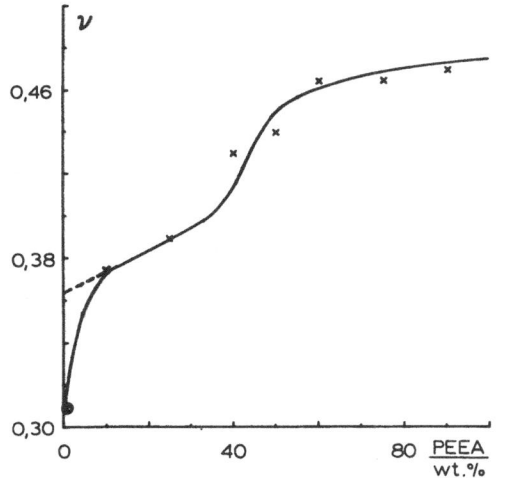

Fig. 20. Dependence of Poisson's ratio ν on composition of a PA 12 based alloy with a PEEA-component of 50 wt % oligolaurolactam

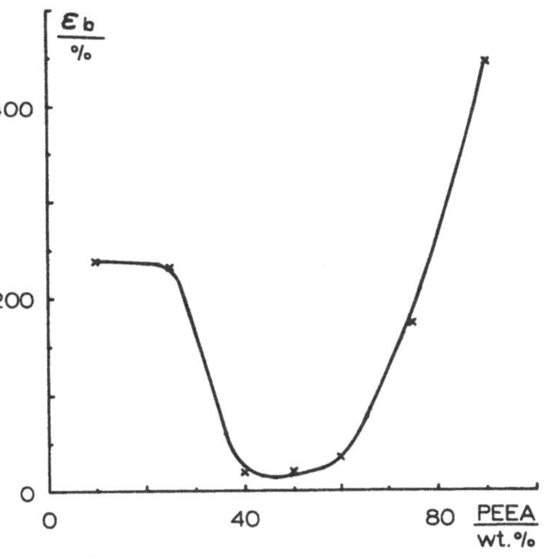

Fig. 22. Dependence of elongation at break on composition of a PA 12 based alloys with a PEEA-component of 50 wt % oligolaurolactam

Such behaviour (a relatively sharp increase in the volume above the yield point) is observed not only in these specific alloys consisting of polyamide 12 and a polyether ester amide containing 50% by weight of hard segments but to a greater or lesser extent also in all alloys of middle composition which have been discussed here.

An overview is given in Figure 21. The slope of the volume/strain curve above the yield point is plotted as

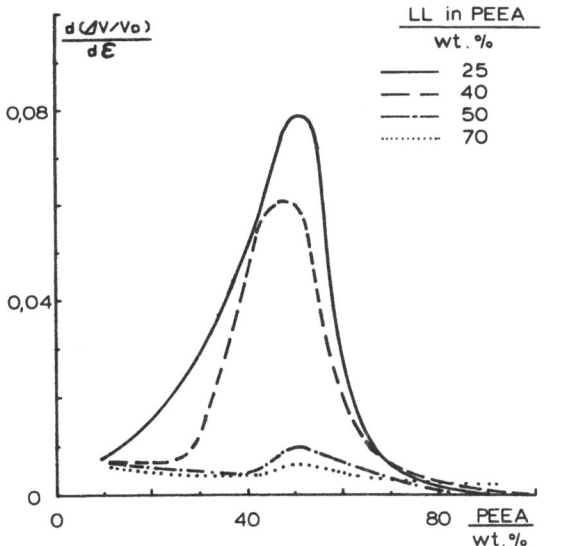

Fig. 21. Dependence of the slope of the volume/strain curve beyond the yield point on the composition of the PA 12 based alloys; parameter: content of oligolaurolactam (LL) in PEEA-components

a function of composition. The parameter noted on the curves is the amount of soft segments in the particular elastomer component. It is found that microcrack formation takes place mainly in alloys of middle composition and is the more pronounced the larger the amount of soft segments in the polyether ester amide.

The slopes of the volume/strain curves are far from unity, which is applicable for pure microcrack formation/crazing, i. e. the deformation above the yield point still takes place predominantly by shear deformation. Nevertheless, microcrack formation in the alloys of middle composition is already so high that the strength properties suffer as a consequence. This can be seen in Figure 22.

The elongation at break for a series of alloys whose polyether ester amide component contains 50% by weight of oligotetrahydrofuran is plotted as a function of composition. It can be seen that the elongation at break has a pronounced minimum exactly in the region in which maximum microcrack formation takes place.

The cause of microcrack formation in the alloys of middle composition is not known with certainty. However, it is obviously associated with the distribution of phases. In the alloys which contain roughly equal amounts, both components are present as a coherent phase. Shear flow processes, of which both components are indeed capable in principle, become more difficult in those cases where the components together form a coherent phase. To date, there are no

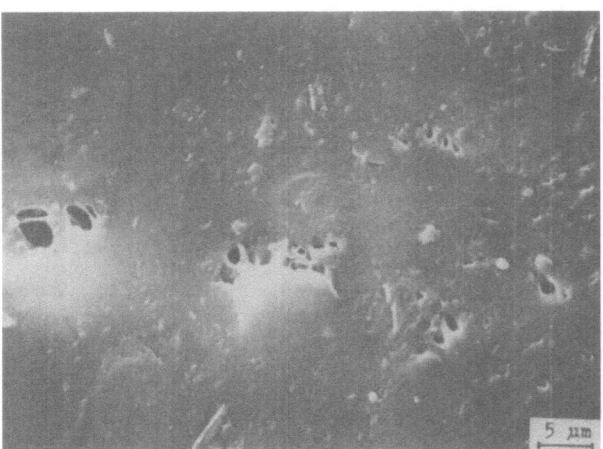

Fig. 23. Scanning electron micrograph of a PA 12 based alloy; composition: 50 wt% PA 12, 50 wt% PEEA with oligolaurolactam

theoretical methods for calculating the distribution of stresses in such complicated structures: however, it must be assumed that in some cases the resulting concentrations of stresses are so high that separation occurs at the phase boundaries in these samples.

In some cases, it is in fact possible to achieve this separation and to localize it.

For example, Figure 23 shows the scanning electron micrograph of a sample which is composed of 50 % by weight of polyamide 12 and 50 % by weight of a polyether ester amide whose oligolaurolactam content in turn is 50 % by weight. It is possible to recognize small cracks and cavities, which appear to have formed in most cases directly at the phase boundaries.

Such cracks or failure phenomena at the phase boundaries are found only in alloys of middle composition. If, on the other hand, one or other of the alloy components predominates, no such effects are detectable in the electron micrographs, in agreement with the measurements obtained by extension dilatometry.

In general, investigations show that the phase boundary strength of the alloys discussed is high. This is evident from the volume/strain behaviour of the systems in which one of the two components is present as the disperse phase. If, on the other hand, both components together form the continuous phase, additional factors play a role.

References

1. Lexikon der Physik, Bd 2 (1969) Franksche Verlagshandlung, W Keller u Co, Stuttgart
2. DKI-Mitteilungen aus dem Deutschen Kunststoffinstitut, Darmstadt (1984) Nr 40, p 11
3. Gude A (1970) Kunststoff-Rundschau 1:6
4. Gude A, Scholten H (1974) Kunststoff-Rundschau 21:77
5. Kraft R (1968) Chem Ind XX:783
6. Mumcu S, Burzin K, Feldmann R, Feinauer R (1978) Angew Makromol Chem 74:49
7. Bornschlegel E, Goldbach G, Meyer K (1985) Progr Coll & Polym Sci 71:119
8. Flory PI (1949) J Chem Phys 17:223
9. Wegner G, Li-Lan Zhu, Lieser G (1981) Makromol Chem 182:231
10. Lohmar J, Angew Makromol Chem, to be published
11. Coumans WJ, Heikens D (1980) Polymer 21:957
12. Coumans WJ, Heikens D, Sjoerdsma SD (1980) Polymer 21:103
13. Bucknall CB (1977) Toughened Plastics, Applied Science Publishers Ltd
14. Bucknall CB, Clayton C (1972) J Mat Sci 7:202

Received January 22, 1986;
accepted January 24, 1986

Authors' address:

Dr. Günther Goldbach
Hüls Aktiengesellschaft
FEA 5, PB 15
Postfach 1320
D-4370 Marl, F.R.G.

Polymer alloys – polymer Blends
Their structure and properties*)

G. Illing

Dr. Illing GmbH + Co. KG, Groß-Umstadt, F.R.G.

The modification of homopolymers with other macromolecular organic substances frequently enables a surprising adaptation to the most diversified practical requirements. Here, the compatibility of the various components and thus the formation of mono-phase to multiphase systems plays a decisive role in influencing the physical properties.

An essential characteristic of this group of substances is the multiphasicity. When no chemical bonds exist between the various phases, it is the phenomenon of the physical mixture, i. e. a blend. In the case of multiphasic systems with genuine chemical bonds between the various phases, i. e. in the presence of graft block polymers, one deals, of course with a completely

different product group, the polymer alloys. Under this designation is understood a macromolecular multi-component system consisting of a continuous phase with the main component A, a discontinuous, embedded phase B and of graft copolymers which consist of segments of the components A and B which are primarily present at the interface between A and B.

The structure of the mentioned types of compounds may easily be recognized on electron microscope images of surfaces of fragments of such multi-component systems. In polymer mixtures or polymer blends, polymer particles of the polymer component (B) are embedded in the continuous phase of the main component (A) without forming any bond whatsoever between the various components at the phase interface. This can easily be seen on a SEM image of the surface of a fragment of a physical mixture consisting of 90 % polycaprolactam and 10 % polyethylene (Fig. 1). In

---

*) Lecture presented during the 32nd Annual Meeting of the Kolloid-Gesellschaft, Berlin October 2–4, 1985.

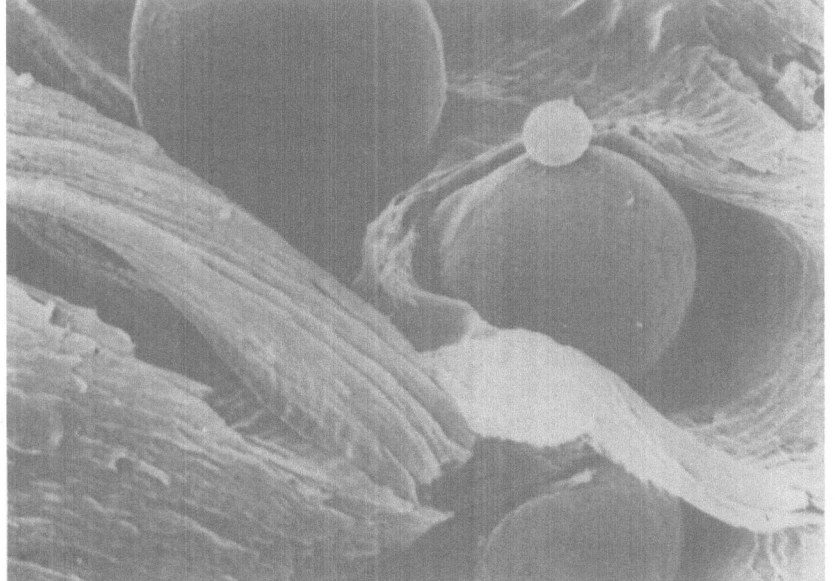

Fig. 1. SEM image of the surface of a fragment of a physical mixture (polymer blend), consisting of 90 % polycaprolactam and 10 % polyethylene

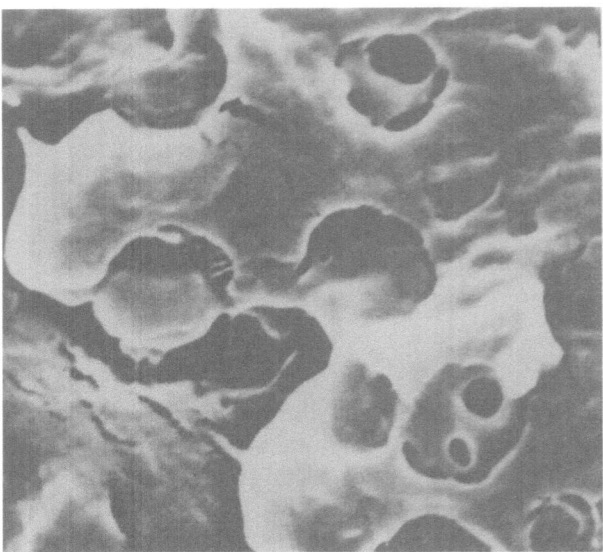

Fig. 2. SEM images of surface of fragments of polymer alloys

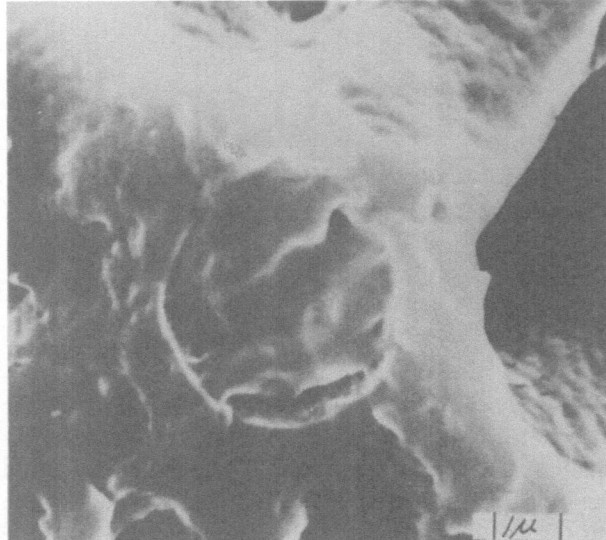

Fig. 4. SEM image of surface of fragments of a polycaprolactam-polyolefin alloy

polymer alloys on the other hand a bond exists between the embedded polymer particles and the surrounding polymer matrix, as shown clearly in Figures 2, 3 and 4. To some extent the interfaces are washed out, and venous bonds can partly be discerned between the embedded particles and the surrounding continuous phase.

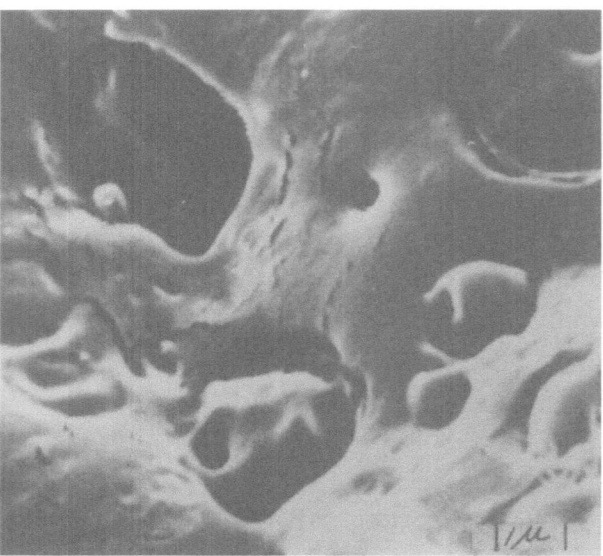

Fig. 3. SEM images of surface of fragments of polymer alloys

The surface electron microscope images (SEM) were taken by Professor Heikens of TU Eindhoven. Injection molded parts of both polymer blends and polymer alloys were placed in liquid nitrogen and thus cooled to −196 °C. Subsequently the samples were fragmented and placed under the electron microscope.

There are several other methods to distinguish polymer blends from polymer alloys, for instance the behavior upon dissolution. It is based on the effect of the formation of stable colloidal solutions, when two different and incompatible polymers are brought into a solvent in which the one polymer component is soluble and the other one isoluble. In this case, stable colloidal solutions are formed due to the dispersing influence of graft polymers which act as emulsifiers. In the case of polymer mixtures (polymer blends) it is possible to separate the various polymer components. Whereas the one polymer component dissolves readily in the solvent, the other insoluble polymer component forms a separate layer (Fig. 5).

Figure 5 shows a colloidal polyamide-polyolefin mixture on the left, and on the right a polyamide-polyolefin mixture, in which the polyamide component is clearly dissolved and the polyolefin component floats undissolved on top. On the basis of investigations by Molau [1—3], this test is also called the Molau test. According to Molau, the emulsifying effect of the graft copolymers depends on the length of the graft side chains. Only above a certain chain length the

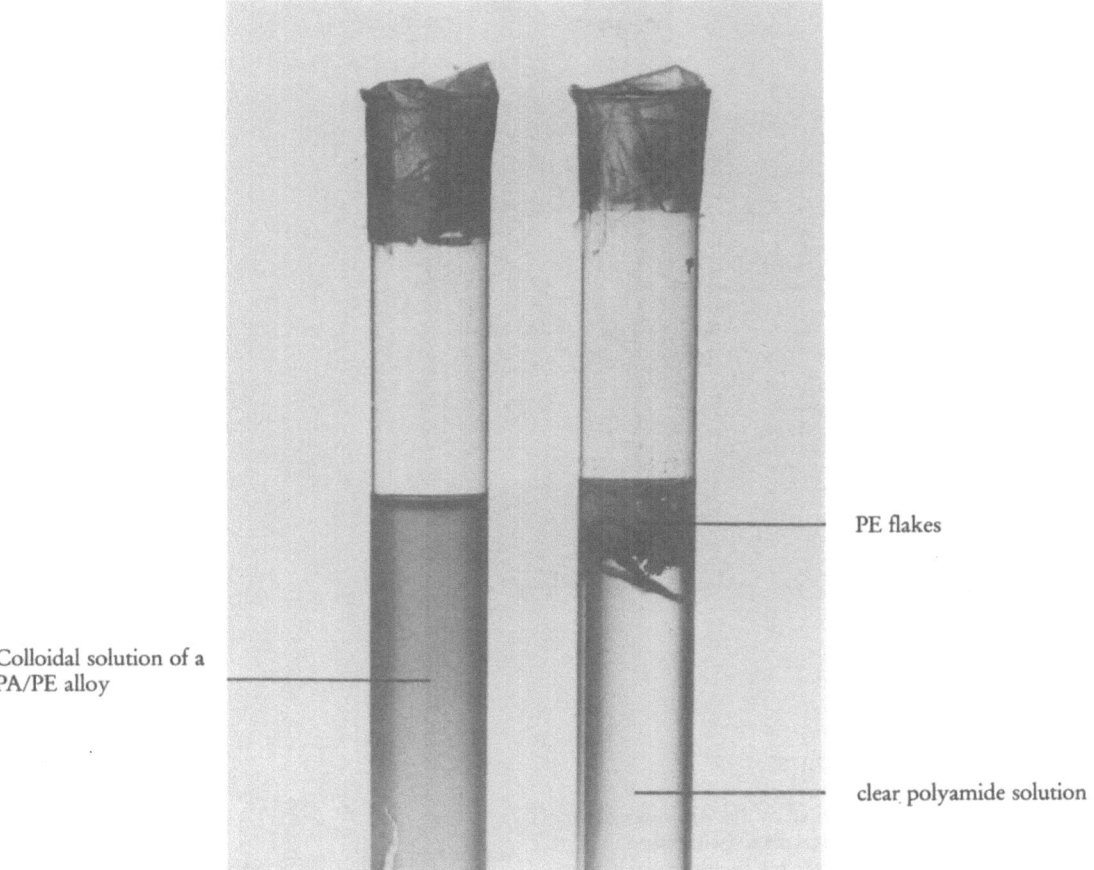

Colloidal solution of a
PA/PE alloy

PE flakes

clear polyamide solution

Fig. 5. Left: a colloidal solution of a polyamidepolyolefin alloy in formic acid. Right: a solution of a polymer blend of polyamide and polyethylene. Polyethylene floats on top of the clear polyamide solution

emulsion stabilizing effect become clearly evident. The graft copolymers are located in the interface of the emulsion droplets, so that the respective chain component *A* is embedded in phase *A*, and the chain component *B* in phase *B*.

There are, of course, distinct differences in the physical properties polymer alloys and polymer blends. It is easy to understand that, for instance, the tensile strength of a polymer blend, where a polymer component *B* is embedded without any chemical bond in the continuous phase of the component *A*, must be considerably lower than that of a polymer alloy, as in the latter genuine chemical bonds exist at the phase interface, and accordingly there is a larger effective cross section.

Finally there are interrelationships between the structure and the apparent properties. Due to their

high compatibility at the phase interface, polymer alloys have a wide range of miscibility, frequently even miscibility in any proportion, a uniform and smooth surface, and optimal uniformity of the properties, even under varying processing conditions (Fig. 6). In this respect, they are very similar to homopolymers or copolymers. Mixtures and blends, on the other hand, do not have these characteristics, or only to a very minor extent, although this does not mean that they are not perfectly suitable for certain applications, provided that specific conditions are maintained during production and processing. Due to their diversified and relatively simple variation potential inherent in these multicomponent systems, these organic materials have established and basic conditions for many new fields of application.

Fig. 6. Formed parts made of a PA 6/PE alloy (right) and a PA 6/PE mixture (blend) (left)

References

1. Molau GE (1965) J Polym Sci A 3:1267,
   (1970) Koll Z, Z Polym 238:493
2. Cherdron H (1978) Paper presented at the 6th International
   Macromolecular Symposium, Interlaken, Switzerland
3. Riess G (1978) Paper presented at the 6th International Macromolecular Symposium, Interlaken, Switzerland
4. Illing G (1981) Angew Makromol Chem 95
5. Illing G, Arndt J (1982) Kunststoffe 72:87–90

Received November 28, 1985;
accepted May 18, 1986

Author's address:

Dr. Illing GmbH & Co KG
Postfach 1150
D-6114 Groß-Umstadt, F.R.G.

Progress in Colloid & Polymer Science

Progr Colloid & Polymer Sci 72:101–105 (1986)

Investigation on crystallization and melting behaviour of linear low density polyethylene (LLDPE)*)

H. Springer, A. Hengse, J. Höhne, A. Schich, and G. Hinrichsen

Technische Universität Berlin, Institut für Nichtmetallische Werkstoffe, Polymerphysik, Berlin, F.R.G.

Abstract: Two commercial types of LLDPEs, Stamylex 1048 and Dowlex 2045, are subjected to definite annealing procedures. DTA measurements are performed during heating and cooling. In the melting as well as in the crystallization interval three distinct processes can be observed. These can be attached to the melting or crystallization of three fractions crystallizing in a different manner.

Key words: LLDPE, DTA, melting, crystallization, segregation.

1. Introduction

Linear low density polyethylenes (LLDPEs) are linear polyethylenes of low density (about 0.92 – 0.93 g/cm³) which are copolymerized in low or high pressure processes using catalysts. The comonomers are ethylene and α-olefins (propene, 1-butene, 1-hexene and/or 1-octene). The molecular structure of LLDPE differs from that of LDPE by the lack of long chain branching, so that some properties of LLDPE rather correspond to those of HDPE, as for example the high melting point and the rheological properties. LLDPE shows some superior qualities in comparison to conventional LDPE: greater tensile and tear strength as well as higher environmental stress crack resistance [1, 2]. This leads to better economy of this material, greater ductility and operating reliability.

It suggests itself that the particular chemical structure of LLDPE also causes special morphological superstructures, which on their part are directly responsible for modifications of the physical and technical behaviour of LLDPE compared to that of LDPE. The aim of this work is to obtain some insight into the crystallization and melting behaviour of LLDPE. It has already been observed that the melting endotherms show several peaks, depending on thermal pretreatment and comonomer content [3].

In this paper we deal with two commercial LLDPEs (Stamylex 1048 and Dowlex 2045). These were subjected to different thermal treatments and then mainly investigated by DTA.

2. Experimental

Two solution polymerized LLDPEs are used in this work. Table 1 gives a rough characterization of these materials.

DTA measurements were performed on a Mettler thermal analyzer. Heating rate was 10 K/min if not otherwise stated. As is well known, thermograms strongly reflect the previous thermal history of the samples. Therefore all samples were in a first stage heated up to 443 K for 20 min.

Table 1. Characterization of materials

Name of LLDPE	Stamylex 1048	Dowlex 2045
Producer	DSM	Dow Chemical
Density [g/cm³]	0.922	0.920
M.F.I. (190, 2.16)	4.5	1.0
Comonomer	1-octene/1-butene	1-octene
Comonomer content (mol %)	1.5/3.2[a])	1.7[a])
Number of branches per 1000 C-atoms	8/16[a])	9[a]) 11,5[b])

---

*) Lecture presented during the 32nd Annual Meeting of the Kolloid-Gesellschaft, Berlin October 2–4, 1985.

[a]) estimated from density and type of comonomer; [b]) according to [1] (determined by C^{13}-NMR)

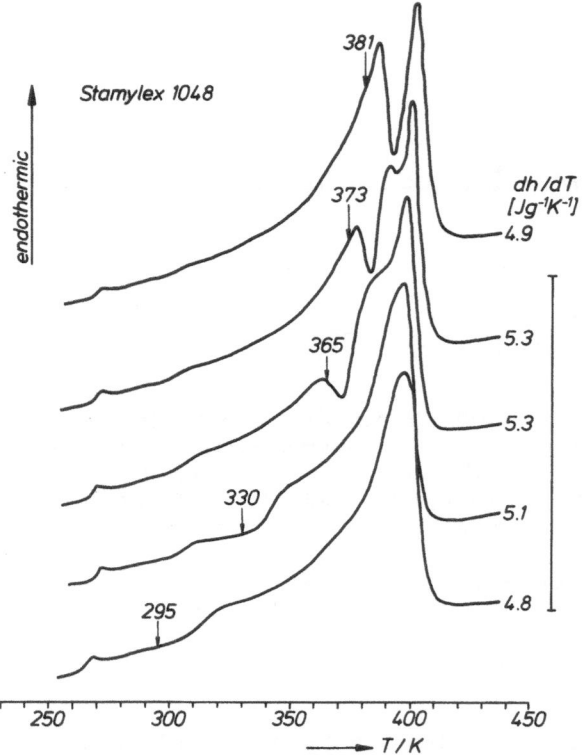

Fig. 1. DTA curves of Stamylex 1048 (series I). The annealing temperatures are indicated by arrows. The ordinate scale is slightly different for each DTA curve and is given by the value at the right end of the curves and the scale bar

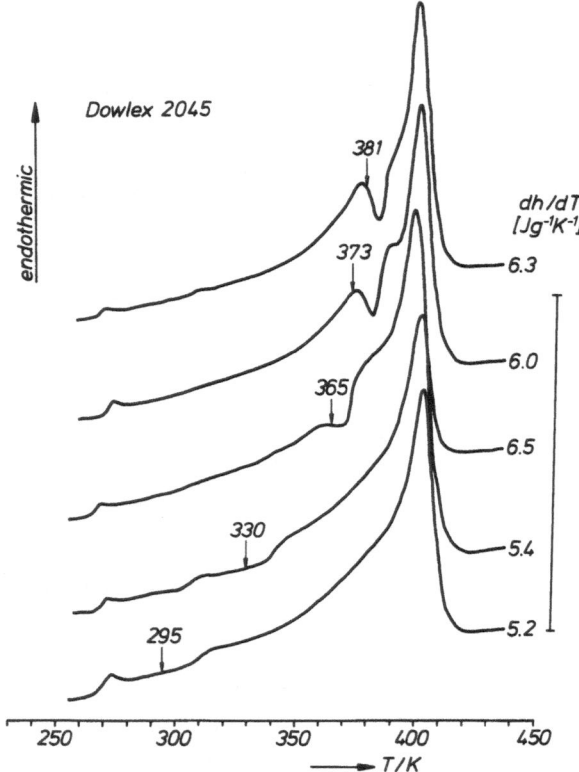

Fig. 2. DTA curves of Dowlex 2045 (series I). The annealing temperatures are indicated by arrows. Ordinate scale as in Figure 1

Samples (about 1 cm³ on the bottom of a test tube) of the main series (series I) were subsequently put into a thermostat with a preselected temperature T_a (295, 330, 365, 373 or 381 K). The average cooling rate from 443 K to T_a was estimated to be about 10 K/min. After annealing at T_a for 24 h the samples were quenched under running water to ambient temperature.

For special investigations and in order to achieve more defined cooling rates, some samples of Stamylex 1048 were also pretreated in the DTA apparatus itself. The first step was as before: annealing at 443 K for 20 min. Thereafter the samples were cooled down to $T_a =$ 373 K with a preset rate and annealed for a defined period. For one series (series II) the annealing period was kept constant at 1 h and the cooling rates were varied: 30, 10, 1, and 0.2 K/min. For a second series (series III) the cooling rate was kept constant and the annealing period was varied: 10 min, 1 h, and 15 h. After annealing the samples of both series were cooled down to ambient temperature with 20 K/min.

The measurement of thermal expansion was carried out on a Perkin Elmer TMA. The TMA specimens were cut from blocks prepared from granules molten at 427 K for 1 h, and after cooling to room temperature annealed at 383 K for 24 h and quenched in ice water. Polarization microscopy was performed on a Leitz polarizing microscope using thin sections of about 30 μm thickness.

3. Results and discussion

Figures 1 and 2 show DTA curves of samples of series I for Stamylex 1048 and Dowlex 2045 respectively. Depending on the annealing temperature up to three peaks can be distinguished in the melting region. Furthermore one recognizes steps at about 270 and 310 K. Analogous steps are found in thermomechanical analysis shown in Figure 3 for Stamylex 1048; an additional step appears at 220 K. The step near 310 K can be attributed to a molecular process in the interior of the crystals. The transitions at about 220 and 270 K are attached to motions of sterically hindered chains within the noncrystalline regions [4–6].

In the following the course of the thermograms within the melting intervals will be discussed in more detail. For the annealing temperature $T_a = 373$ K three peaks can be clearly distinguished. We assign the corresponding peak temperatures as T_1, T_2 and T_3 with increasing temperature.

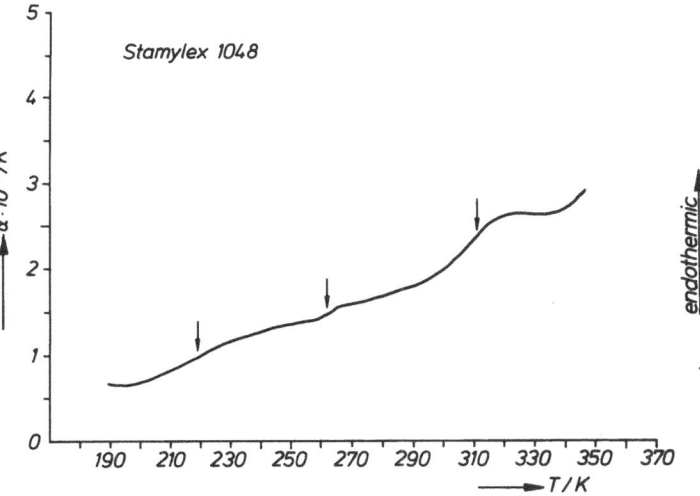

Fig. 3. Thermal expansion coefficient for Stamylex 1048 as a function of temperature. The steps are indicated by arrows

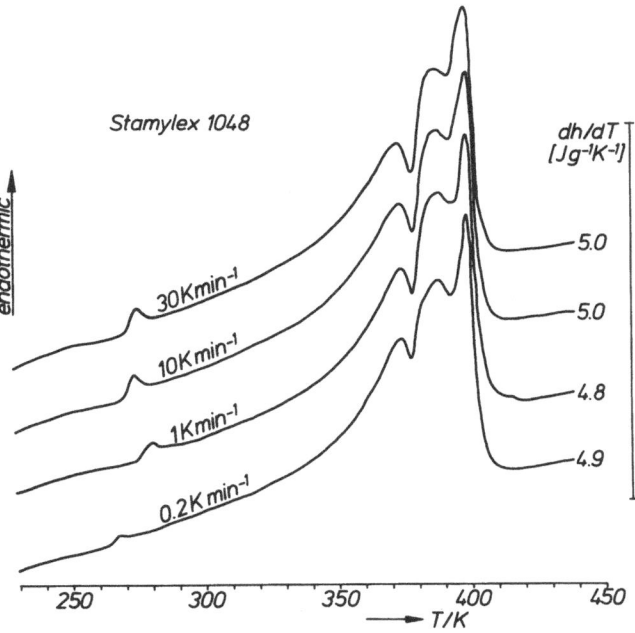

Fig. 5. DTA curves of Stamylex 1048 (series II). Parameter: rate of cooling from 443 to 373 K

For Dowlex T_3 is situated at about 403 K, whereas for Stamylex it takes values between 397 and 402 K, the higher ones being observed for samples annealed at higher temperatures. The T_3-peak appears in a temperature range, in which also linear polyethylene (HDPE) shows its melting maximum. Therefore it seems to be reasonable to attribute the T_3-peak to a fraction of the LLDPE chains, which are able to crystallize in the same manner as HDPE chains. In the

DTA-cooling curves (Fig. 4) a peak corresponding to the T_3-maximum is observed. This high temperature crystallization peak strongly depends on the cooling rate. As can be gathered from Figure 5, the T_3-peak of samples cooled more slowly from 443 to 373 K appears more pronounced. This effect may be due to the fact that the corresponding crystallization process takes place in a narrower temperature interval. Shortening of the annealing period has no influence on T_3 (Fig. 6).

The splitting up of the melting regime into three peaks is clearly observed only on specimens annealed at $T_a = 373$ K. The separation becomes still more marked, if the heating rate is decreased (Fig. 7). The positions of T_a as indicated in Figures 1 and 2 have to be discussed. As shown for example in [7], one would expect T_a to lie at the minimum between T_1 and T_2. The observed discrepancy has the following reasons: firstly, the abscissa of the DTA curves gives the temperature of the reference cell. A plot against the sample temperature, obtained by correcting for the height of the DTA signal would shift the DTA peak by nearly 2 K to lower temperatures. Secondly, the peak position depends on the heating rate: lowering the heating rate from 10 to 1 K/min shifts the peak position about 3 K to the left (see Fig. 7). Thirdly the annealing temperature

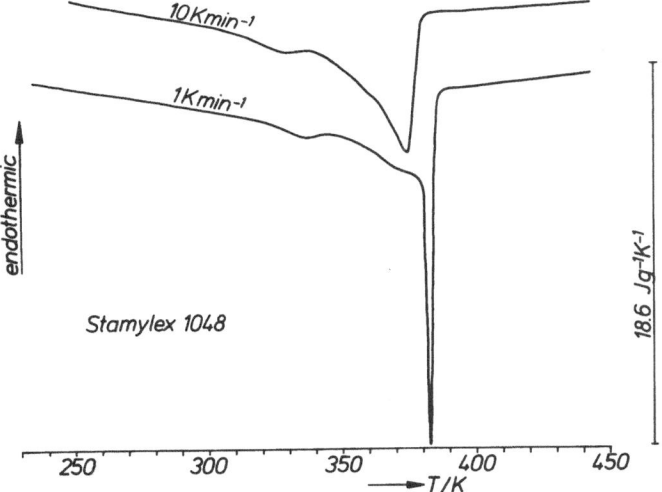

Fig. 4. Crystallization curves of Stamylex 1048 for two different cooling rates

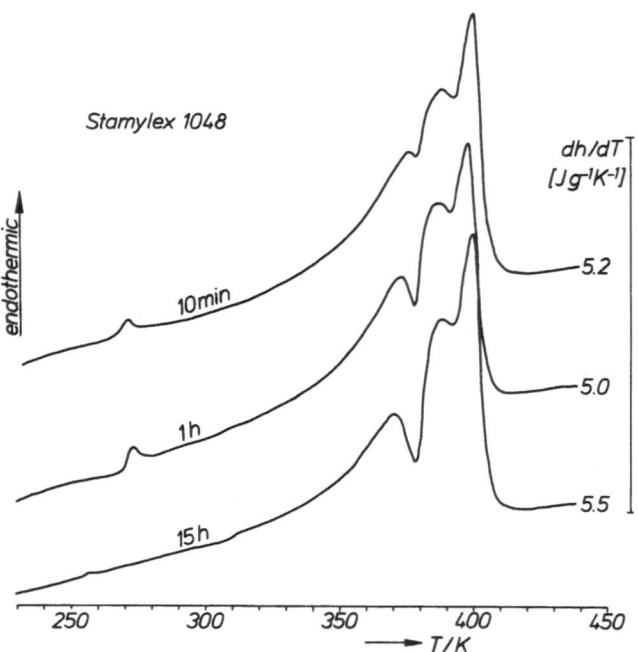

Fig. 6. DTA curves of Stamylex 1048 (series III). Parameter: annealing period. Ordinate scale as in Figure 1

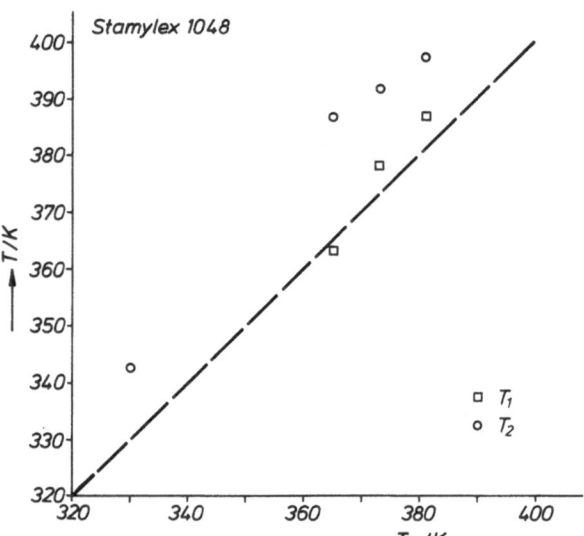

Fig. 8. Characteristic temperatures T_1 and T_2 for Stamylex 1048 plotted against annealing temperature T_a

reached in the sample may be incorrect by 2 K. Thus the true annealing temperature may approach to the temperature of the minimum between T_1 and T_2.

For annealing temperatures different from 373 K the DTA traces show two peaks, two peaks and one shoulder or one peak and one shoulder. If one plots the positions of the peaks (leaving aside T_3) and shoulders

against the annealing temperature (Figs. 8 and 9) these can be divided into two classes: for one the temperatures lie about 10 K above T_a, for the other one they are nearly equal to T_a. We denote the temperatures of the two classes with T_2 and T_1 respectively according to the above assignment of the temperatures for the sample of series I annealed at 373 K.

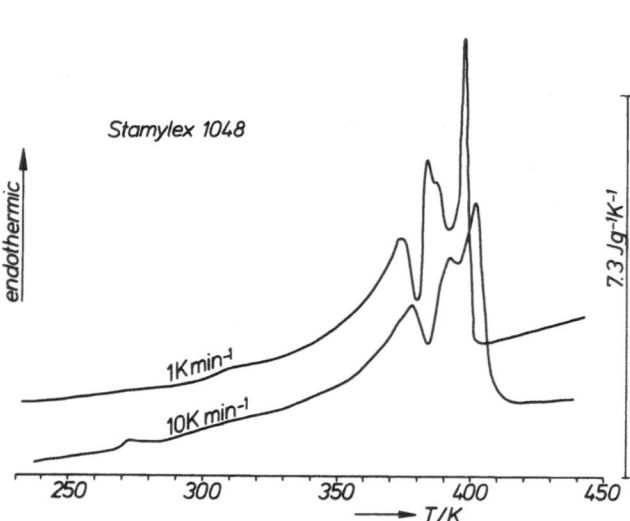

Fig. 7. Dependence of the shape of DTA curves on heating rate for Stamylex 1048 (series I, $T_a = 373$ K)

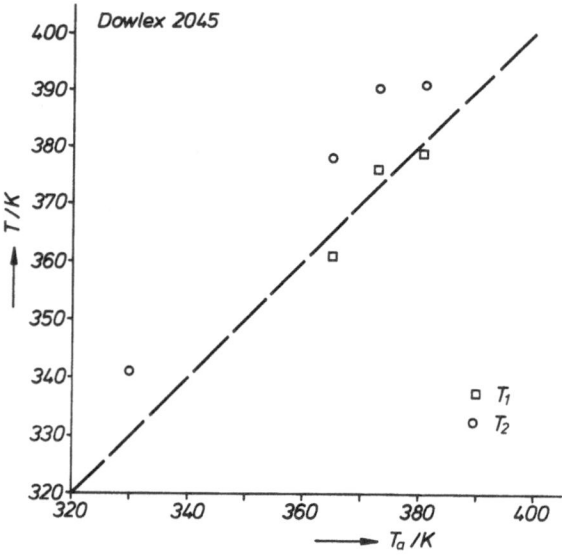

Fig. 9. Characteristic temperatures T_1 and T_2 for Dowlex 2045 plotted against annealing temperature T_a

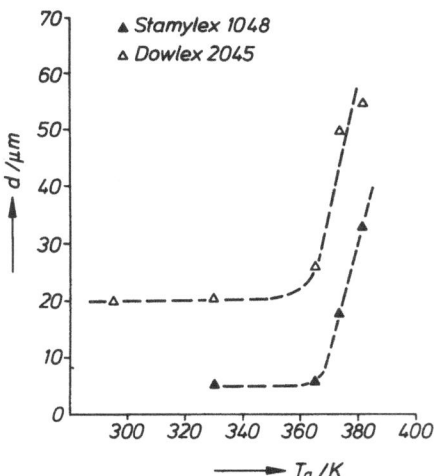

Fig. 10. Average spherulite diameter as a function of annealing temperature T_a

The melting process at temperature T_2 obviously arises from a portion of material which has been recrystallized at T_a to crystallites of higher perfection. This leads to the well known phenomenon of the so-called "annealing peak" (see for example [8]). The material which participates in the recrystallization process may originally have been crystallized in a temperature interval corresponding to the broad shoulder following the high temperature peak in the crystallization curve.

The maximum of the melting curves close to the annealing temperature must be caused by a fraction of crystallites, the chains of which do not crystallize at or above T_a. Indeed also in the crystallization curves (Fig. 4) a second exothermic peak is found at low temperatures. This leads to the assumption that a low viscosity fraction exists within the LLDPE samples which is able to segregate during annealing at adequate temperatures (cf. [3]). This assumption recently has been confirmed in [9], where intermolecular variation of copolymer statistic in LLDPE has been proven. The segregation process is strongly time dependent as can be seen from the thermograms of series III samples (Fig. 6). The appearance of analogous peaks at relatively high annealing temperatures is reported and extensively discussed for LDPE in [7, 8, 10] and for HDPE in [11]. Thus peaks like the T_1- and T_2-peaks are found for LDPE and HDPE. Against that, multiple peaks are found for LLDPE [3] and LDPE/HDPE-blends [12, 13] simultaneously after annealing only at one temperature.

The superstructure of the LLDPE samples of series I is characterized by space-filling spherulites exhibiting twisted lamellae. The average diameter of these depends on the annealing temperature (Fig. 10). According to [10] the annealing temperature is not identical with the temperature of spherulite growth, and at the lower T_a-values the temperature range of spherulite growth is nearly independent of T_a, which is reflected in the constancy of the spherulite diameter.

4. Conclusion

DTA curves of the investigated commercial LLDPEs show during the melting process of annealed samples as well as during the crystallization process of samples cooled down from 443 K up to three clearly separable regimes. Three different fractions of the materials may cause this behaviour: firstly, an especially well crystallizing fraction gives rise to the high temperature peak, secondly, the medium regime originates in slower crystallizing chains, which recrystallize during annealing, and thirdly, a low viscosity fraction segregates during annealing and results in the low temperature peak.

Additional work has to be done to elucidate further the chemical and physical nature of these three fractions.

References

1. Maddams WF, Woolmington J (1985) Makromol Chem 186:1665
2. Bork S (1984) Kunststoffe 74:474
3. Schouterden P, Riekel C, Koch M, Groeninckx G, Reynaers H (1985) Polym Bull 13:533
4. Kakizaki M, Kakudate T, Hideshima T (1985) J Polym Sci Polym Phys Ed 23:809
5. Saini DR, Shenoy AV (1985) Polym Commun 26:50
6. Phillips PJ, Wilkes GL, Delf BW, Stein RS (1971) J Polym Sci A-2, 9:499
7. Strobl GR, Schneider MJ, Voigt-Martin IG (1980) J Polym Sci Polym Phys Ed 18:1361
8. Pakula T (1982) Makromol Chem 183:1577
9. Mathot VBF, Schoffeleers HM, Brands AMG, Pijpers MFJ (1985) Proc 17 Europhys Conf Macromol Phys 1985
10. Pakula T (1982) Polymer 23:1300
11. Yadav YS, Jain PC, Nanda VS (1985) Thermochim Acta 84:141
12. Clampitt HB (1963) Anal Chem 35:577
13. Clampitt HB (1964) Americ Chem Soc Polym Prepr 5:354

Received December 2, 1985;
accepted April 10, 1986

Authors' address:

Dr. H. Springer
Institut für Nichtmetallische Werkstoffe, Polymerphysik
Englische Straße 20
D-1000 Berlin, F.R.G.

Polymer solutions with threshold values of shear stress and shear rate in the flow curve*)

H. Bauer and G. Meerlender

Physikalisch-Technische Bundesanstalt, Braunschweig, F.R.G.

Abstract: The rheological behavior of solutions of a styrene-isoprene diblock copolymer in C_8 and C_{11} n-paraffins (polymer content 5 % to 7 %) and in mixtures of n-paraffins with cyclohexane are reported. The flow curves as measured with rotational viscometers show two pronounced discontinuities at finite shear stress and shear rate: a threshold shear stress (yield point) around 10 Pa and a threshold shear rate for the onset of an abrupt shear thickening (order of magnitude 100 s^{-1}). The flow characteristics are reproducible and time independent. The influences of solvent composition, polymer content and temperature are discussed. A linear model and a possible molecular arrangement to this model are proposed.

Key words: Polystyrene-polyisoprene copolymer, diblock copolymer solution, threshold shear rate, plastic flow, shear thickening.

1. Introduction

Probably the best known example of a discontinuity in the flow curve at finite values of shear stress τ and shear rate $\dot{\gamma}$ is the yield point of a plastic material. In the simplest idealization this is described by the Bingham model. The material undergoes flow when a threshold value of the shear stress is exceeded. The complementary behavior, the occurrence of a threshold value of the shear rate, however, is observed only in few systems. Moreover, time effects (i. e. change of the rheological properties with time during the flow and caused by it) often play a role, with the consequence that, depending on the experimental arrangement, a threshold shear rate may be only simulated. For example, antithixotropic behavior was observed with certain polymer solutions (so-called 'antimisting cerosene') when rotational viscometers were used to measure the flow characteristics [1]. If, however, flow curves were calculated from measurements with capillary viscometers, an apparent threshold value of the shear rate was observed in the flow curve of the same material [2].

One example of the existence of a true threshold shear rate independent of time effects is the system of polyvinylalcohol-borate complexes in aqueous solution [3].

In the present work the rheological behavior of polymer solutions is described showing both types of discontinuities in the flow curve, a threshold shear stress as well as a threshold shear rate, independent of time effects.

2. Experimental

The polymer used in this investigation is a linear diblock copolymer consisting of styrene and isoprene (SI), the latter being hydrogenated after the polymer has been formed. The average molecular mass is about 100 000, the mass distribution is characterized by a ratio of mass average to number average of 1.4. The mass fraction of the styrene block in the polymer molecule is 37 %. This polymer is dissolved in n-octane, n-undecane and in mixtures of these paraffins with cyclohexane by gently stirring it, usually at a temperature of 60 °C. The solutions are then stored at room temperature until the rheological properties show no change with time, which is usually achieved after four weeks. All samples are homogeneous and optically clear transparent gels.

Flow curves were measured in rotational viscometers using concentric cylinder systems according to DIN 53019, double gap cylinder systems, and in some cases, also cone and plate arrangements. For the evaluation of the results the concept of representative values τ_{rep} and $\dot{\gamma}_{rep}$ for shear stress and shear rate is used [4].

---

*) Lecture presented during the 32nd Annual Meeting of the Kolloid-Gesellschaft, Berlin October 2–4, 1985.

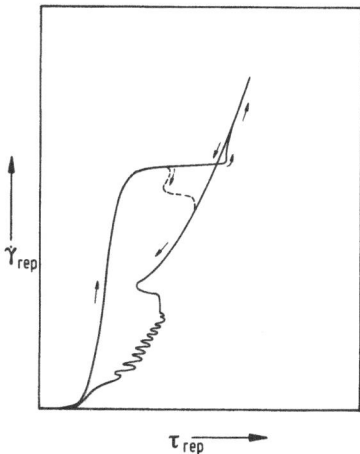

Fig. 1. Typical flow curve with yield point and shear thickening plateau

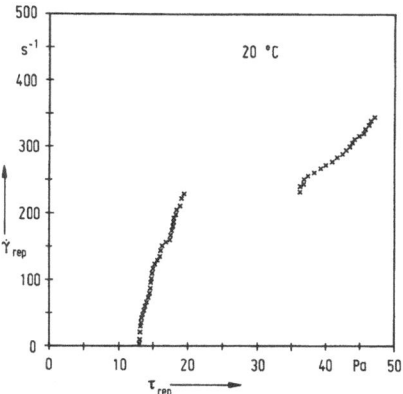

Fig. 2. Flow curve for a solution of 6% polymer (SI) in a mixture of n-octane / 20% cyclohexane, measured in a double-gap cylinder system by increasing the shear rate stepwise and waiting for viscometric flow

3. Results

A typical example of the unusual type of flow curve observed with these polymer solutions is given in Figure 1. At low shear rates the flow curve is very similar to that of a Bingham body. The shear stress at the yield point is of the order of 10 Pa, depending mainly on the polymer concentration. As the shear rate is increased, an abrupt shear thickening occurs, causing a transition to a different rheological state of the sample. The shear stress increases by more than a factor of two within a very narrow interval of shear rate. We will call this the shear thickening plateau. The plateau shear rate is of the order of 100 s^{-1} and depends mainly on temperature and solvent composition. For shear rates above this transition, the flow curve is almost a straight line with only a slight tendency to a shear thinning behavior. The differential viscosity $d\tau/d\dot{\gamma}$ in this state is significantly higher than in the initial state at low shear rates. Both the lower and the upper branch of the flow curve are reproducible and reversible with respect to the increase and decrease of the shear rate unless the transition range is passed downwards where the plateau of shear thickening is not accessible. In this range, instabilities in the flow field sometimes occur, as indicated by the oscillations in the flow curve in Figure 1. The lower branch of the curve, however, is restored immediately when the shear rate is brought to zero, even by a sudden stop when proceding from a higher value (typically 450 s^{-1}). The lower branch of the flow curve (according to the low viscosity state of the sample) is reversible only as long as the shear rate stays below the critical transition range. If the shear rate is reduced from a value in the plateau region, a transition to the high viscosity state of the sample will occur. This is indicated by the dashed line in Figure 1. At the upper end of the plateau the inertia of the rheometer and the recording system produces an overshoot in the registered flow curve which depends on the speed of shear rate change. This is also the reason for the 'tail' of the flow curve at low shear rates, which is not observed when the flow curve is recorded more slowly or point by point under conditions of true viscometric flow. The occurrence of the shear thickening plateau is not caused by wall slip. It is also independent of the geo-

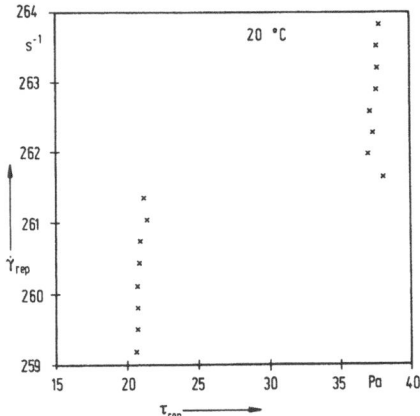

Fig. 3. Threshold shear rate for the shear thickening of a solution of 6% polymer (SI) in a mixture of n-octane / 20% cyclohexane, measured in a double-gap cylinder system by increasing the shear rate stepwise and waiting for viscometric flow; smaller steps than Figure 2.

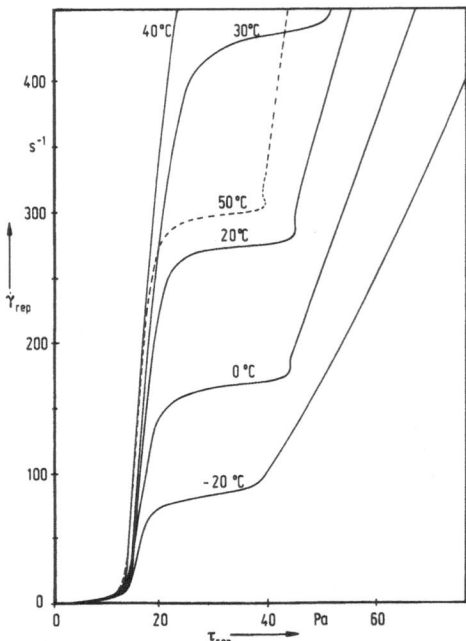

Fig. 4. Flow curves for a solution of 6 % polymer (SI) in a mixture of n-octane / 20 % cyclohexane at different temperatures

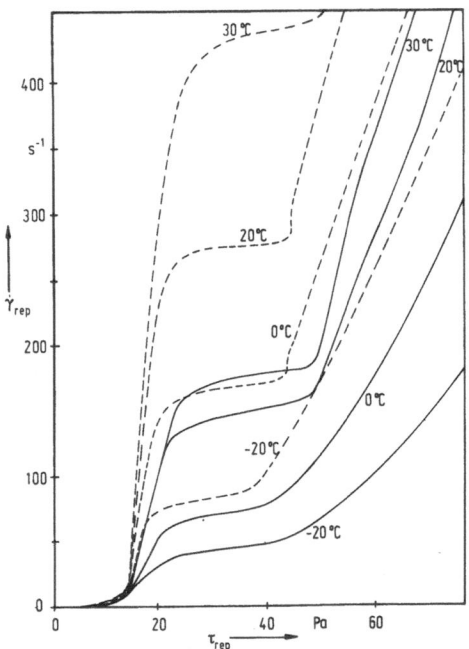

Fig. 5. Influence of the paraffinic component of the solvent on the flow curve at different temperatures for solutions of 6 % polymer (SI) in a mixture of n-paraffin / 20 % cyclohexane. Solid lines: n-undecane, dashed lines: n-octane

metry of the measuring system. When the flow curve is registered point by point under conditions of true viscometric flow, an extremely sharp shear thickening transition can be observed. This is demonstrated in Figures 2 and 3.

The threshold shear rate for the shear thickening increases slightly when the shear rate change during continuous registration or the shear rate interval in point-by-point measurements is decreased. If the shear thickening were caused by a time effect as in [1], the opposite behavior would be observed.

The threshold shear rate depends strongly on the temperature (Fig. 4). With increasing temperature the shear thickening plateau is shifted to higher shear rates until at a certain temperature a maximum value is observed. In the example shown in Figure 4 this maximum is reached at about 40 °C (outside the range contained in Fig. 4). Up to this temperature the yield point and the shear stress at the onset of the shear thickening plateau do not change much with temperature. By a further increase in temperature the plateau is lowered again, and at the same time, the reproducibility of the flow curve is lost (see dashed curve for 50 °C in Fig. 4). If the temperature is further increased, first the plateau disappeares, then the yield point is lowered until finally, Newtonian behavior is observed.

The chain length of the paraffinic component of the solvent has only little influence on the yield point, but affects the shear thickening plateau, as can be seen from Figure 5. When the cyclohexane content is kept at 20 %, the most pronounced shear thickening plateau is observed for n-octane. An increase in the chain length of the n-paraffin results in a lowering of the threshold shear rate. With a mixture of n-C_{13} to n-C_{17} (main component n-C_{15}) as the paraffinic component of the solvent, shear thickening is no longer observed. On the other hand, with n-hexane (this was investigated with 5 % polymer only) there is still a yield point but no shear thickening. Instead, pure cyclohexane as a solvent for the same amount of polymer results in New-tonian flow behavior. An increase in the polymer content from 5 % to 7 % has only little influence on the threshold value of the shear rate and its temperature dependence, but increases the yield point and the apparent viscosity, as can be seen from Figure 6.

In a previous investigation [5] it was concluded that in order to obtain a shear thickening plateau in the flow curve of this type of polymer solutions, it is necessary to use a mixture consisting of an n-paraffin and a cyclic hydrocarbon as a solvent for the SI-polymer. In

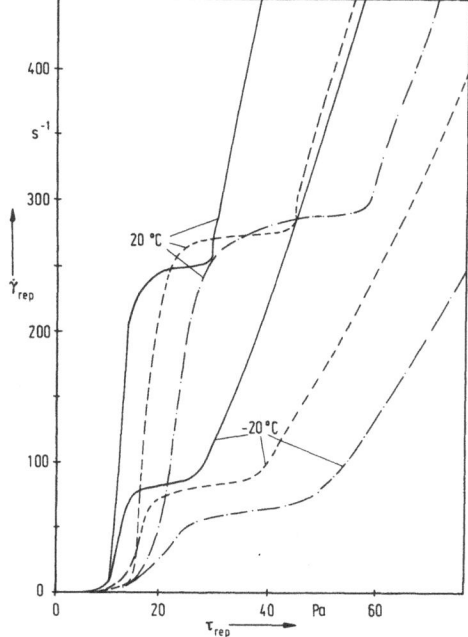

Fig. 6. Influence of the polymer content on the flow curve of SI-polymer dissolved in a mixture of n-octane / 20% cyclohexane. Solid lines: 5% polymer; dashed lines: 6% polymer; dash-dotted lines: 7% polymer

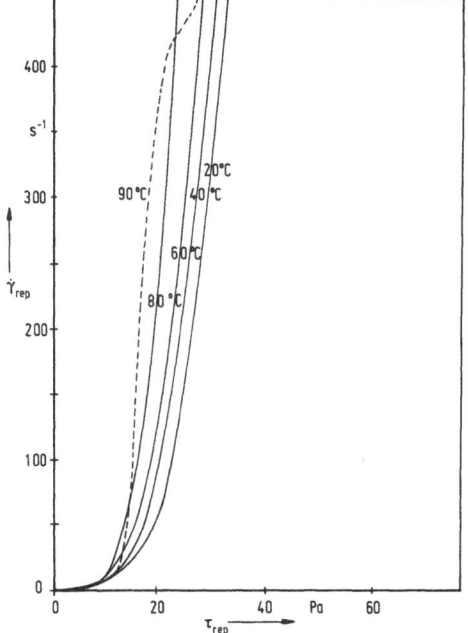

Fig. 7. Flow curves at different temperatures for a solution of 6% polymer (SI) in pure n-undecane. The solution was prepared at 60°C

samples prepared with pure n-paraffins no shear thickening effect was observed at the time. Recent results, however, show that also in this case a shear thickening plateau can be obtained if the sample is heated to a certain temperature. This is demonstrated in Figure 7 and Figure 8. Figure 7 shows the temperature dependence of the flow curve for n-undecane with 6% polymer. The sample was prepared at 60°C. Up to 80°C the normal viscosity temperature behaviour is observed. At 90°C first indications of a shear thickening appear, and at the same time the reproducibility of the flow curve is lost (dashed line in Fig. 7). Measurements at 20°C after each temperature increase show that up to 80°C the sample remains in principle unchanged. The temperature increase from 80°C to 90°C obviously causes an irreversible change in the microstructure of the sample, since after this the flow curves at lower temperatures are completely different (Fig. 8). The viscosity is significantly increased and a shear thickening plateau is observed. At 90°C, again the flow curve is no longer reproducible and at 100°C the shear thickening plateau and the yield point have disappeared. It must be noted that this heat treatment does not change the sample composition.

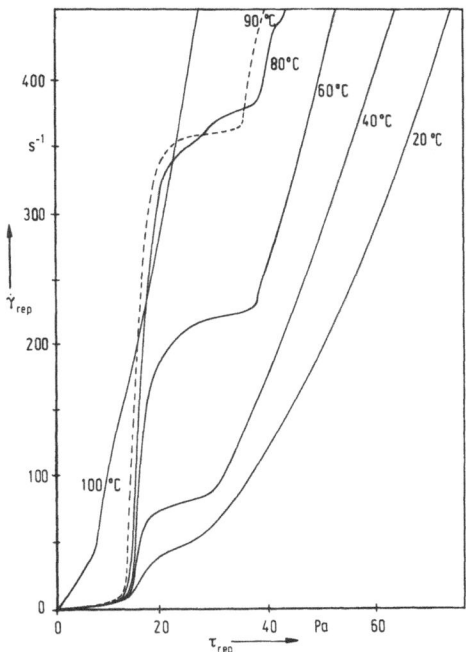

Fig. 8. Flow curves at different temperatures for a solution of 6% polymer in pure n-undecane after heat treatment at 90°C (compare with Fig. 7)

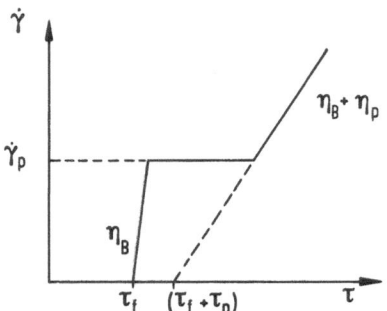

Fig. 9. Idealized flow curve with yield point τ_f and shear thickening plateau $\dot{\gamma}_p$

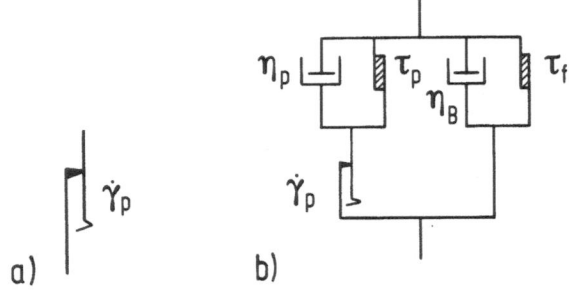

Fig. 10. a) Inverse Saint-Vernant model with

$$\eta = \begin{cases} 0 & \text{for } \dot{\gamma} < \dot{\gamma}_p \\ \infty & \text{for } \dot{\gamma} \geq \dot{\gamma}_p \end{cases}$$

b) Linear network to describe the idealized flow curve in Figure 9

After the sample has been heated to 100 °C the flow curve at 20 °C shows no shear thickening: instead, the upper branch of the flow curve which is almost the same as observed after heating to 90 °C is continued down to the yield point. The shear thickening plateau can be restored by 'annealing' the sample at temperatures up to 80 °C. The higher the temperature, the earlier an equilibrium condition is reached.

4. Description of the flow curve by a linear network

The shape of the new type of flow curve suggests an idealized description in analogy to the Bingham body as shown in Figure 9. The model that describes this flow curve must account for two singularities, one at finite shear stress and one at finite shear rate: the yield shear stress τ_f and the shear rate $\dot{\gamma}_p$ of the shear thickening plateau. There is, however, no way in which this model can be composed of the basic elements: the Newtonian dashpot, the Hookian spring and the Saint-Venant model (ideal plastic body). It is necessary to add an inverse Saint-Venant model for which the viscosity is zero as long as the shear rate is lower than a critical value $\dot{\gamma}_p$. Above this value the viscosity is infinite. A symbolic representation of this element is shown in Figure 10. The system that leads to the flow curve in Figure 9 can be described by two Bingham bodies with different values for the yield shear stress (τ_f, τ_p) and the viscosity (η_B, η_p) which are connected by an inverse Saint-Venant element as shown in Figure 10 b. The algebraic formulation is given by

$$\tau - (\tau_f + h \cdot \tau_p) = (\eta_B + h \cdot \eta_p) \cdot \dot{\gamma}$$

$$h = \begin{cases} 0 & \text{for } \dot{\gamma} < \dot{\gamma}_p \\ 1 & \text{for } \dot{\gamma} \geq \dot{\gamma}_p \end{cases} . \tag{1}$$

Only viscometric flow is discussed and therefore no Hookian springs occur in the model.

5. A Qualitative molecular model

The observation of gelation in a solution of diblock-copolymers in so-called 'selective solvents' which dissolve only one of the components of the polymer, is well known (recent literature on this subject is reviewed in [6]). The existence of a yield point is therefore not unexpected in the system investigated here. In this case the polyisoprene chains are dissolved, whereas the polystyrene blocks are precipitated, forming micelles which are coupled together via the isoprene chains thus forming a gel structure. The flow properties observed suggest a mechanism in which the micelles are responsible for the shear thickening effect.

First, the samples with pure n-paraffin as the solvent are considered (see Figs. 7 and 8). The solutions were prepared at 60 °C. This temperature is not high enough to dissolve the copolymer completely, and gelation begins during the solubilization process. In this case it can be assumed that the polystyrene blocks in the micelles form a relatively rigid structure. During shear flow these micelles will behave like rigid particles, which means that they simply rotate after the yield stress is exceeded. This makes only a small hydrodynamic contribution to the flow resistance. It is in this situation that a yield point but no shear thickening plateau is observed (flow curves at 20 °C up to 80 °C in Fig. 7). If the sample is heated, at a certain temperature solvent will start to penetrate into the micelles (in the example in Fig. 7 at around 90 °C). These swollen micelles are viscoelastic and during shear flow they will not only rotate but they also undergo a periodic deformation which rotates in the opposite direction. This causes an additional energy dissipation inside the micelles and results in the increased viscosity in the upper branch of the flow curve (see Fig. 8). It must be assumed that at low shear rates the swollen micelles also behave like rigid particles. Only when the forces which cause the deformation of the micelles reach a certain threshold value is the viscoelastic deformation of the micelles possible. This threshold value is given by the shear rate at the onset of the shear thickening plateau. It is unlikely that the microstructure of the micelles is changed by this viscoelastic deformation, otherwise the immediate recovery at zero shear could not be understood. The viscosity of the matrix around the micelles decrease with increasing temperature. In order to reach the threshold value for the deformation of the micelles, a higher shear rate is necessary. The shear thickening plateau is therefore shifted to higher shear rates until a temperature is reached where the inner structure of the micelles is changed by temperature and shear. The reproducibility of the flow curve is then also lost. This temperature is about the same as that which is necessary to allow the solvent to penetrate into the micelles. When the temperature has been raised well above this point no shear thickening plateau is found afterwards at lower temperatures. The increased viscosity, however, remains. From this observation it must be concluded that the existence of the threshold shear rate for the deformation of the micelles must be attributed to some kind of ordered molecular structure inside the micelles. Once this structure has been thermally destroyed it takes a long time to restore it. The recovery time is shorter when the mobility of the molecules is increased by 'annealing' the sample.

The influence of cyclohexane (or other cyclic hydrocarbons [5]) as a component of the solvent consists mainly in an increase of the solvent quality. The mechanism described above remains unchanged, only the temperature necessary for the swelling of the micelles and for their destruction is lowered.

This model is also supported by the observation that a sample with pure n-paraffin as the solvent can be brought to a state where a shear thickening occurs not only by heating but with the aid of a good solvent. When the polymer is first dissolved in diethylether, the n-paraffin then added and the diethylether removed by distillation at low temperature, a shear thickening plateau is observed in the flow curve.

Acknowledgement

The authors wish to thank Mrs. A. Weber for her help in performing sample preparation and measurements.

References

1. Peng STJ, Landel RF (1981) J Appl Phys 52:5988
2. Matthys EF (1984) Proc 9th Int Congr Rheol Acapulco, Vol II:117
3. Savins JG (1968) Rheol Acta 7:87
4. Giesekus H, Langer G (1967) Rheol Acta 16:1
5. Bauer H, Meerlender G, Stern P, Rheol Acta (in press)
6. Watanabe H, Kotaka T (1984) Polymer Engng Rev 4:73

Received November 27, 1985;
accepted January 13, 1986

Authors' address:

Dr. G. Meerlender, Dr. H. Bauer
Physikalisch-Technische Bundesanstalt
Bundesallee 100
D-3300 Braunschweig, F.R.G.

Progress in Colloid & Polymer Science   Progr Colloid & Polymer Sci 72:112–118 (1986)

Macromonomers as polymeric intermediates. Synthesis and applications*)

P. Rempp, E. Franta, P. Masson, and P. Lutz

Institut Charles Sadron (CRM-EAHP) (CNRS-ULP), Strasbourg, France

Abstract: The methods developed for the synthesis of macromolecular monomers (macromonomers) are reviewed, with emphasis on those based on ionic "living" polymerization techniques which give access to well-defined species of low polydispersity. The ability of the terminal unsaturation to undergo free radical polymerization (or copolymerization with a vinylic or acrylic monomer) is considered next. Homopolymacromonomers are characterized by an unusually large segment density within the polymer coil. The chief interest of macromonomers is the easy access to graft copolymers they can provide, especially for amphiphilic species constituted of a hydrophobic backbone carrying hydrophilic grafts. Further applications of macromonomers can also be envisioned.

Key words: Macromonomers, ionic polymerization, functionalization at chain end, graft copolymers, branched polymers.

A "macromolecular monomer" is a short polymer chain which carries at one chain end — or at both — a polymerizable feature (a double bond, or a heterocycle capable of undergoing ring-opening polymerization) or even functions able to participate in a polycondensation reaction to yield graft copolymers.

$$R-(CH_2-CH)_n-CH_2-\langle\bigcirc\rangle-CH=CH_2$$
$$|$$
$$C_6H_5$$

$$HO-(CH_2-CH_2-O)_n-CH_2-\underset{O}{\diagdown}$$

$$R'-(CH_2-CH)_n-S-CH-COOH$$
$$\quad\quad | \quad\quad\quad |$$
$$\quad\quad R \quad\quad CH_2-COOH$$

It was necessary to restate this definition, because of frequent misuses of the word macromonomer. Neither telechelic polymers nor mesogenic monomers should be referred to as macromonomers.

Early attempts to synthesize polymer chains fitted with end-standing unsaturations were performed in the 1960s by Bamford [1], Greber [2], Gillman [3] and other authors. The first systematic research in that field was however carried out by the late Ralph Milkovich [4]. In the past ten years macromonomers have received increasing interest in universities as well as in industrial laboratories.

It was shown that these species are versatile intermediates in macromolecular synthesis and that they particularly provide an easy access to graft copolymers [5].

In the first part of this report the preparation of macromonomers will be outlined. The second part will be devoted to the ability of these macromolecular monomers to undergo polymerization. Only macromonomers with an end-standing double bond — by far the most studied ones — will be considered.

---

*) Main Lecture at the Berliner Polymeren Tage 1985.

1. Synthesis of macromonomers

The synthesis of macromonomers requires a careful characterization of the species formed. It has to be established whether every single molecule is fitted with an end-standing double bond. This is usually done by comparison between the number average molecular weight of the sample determined directly by SEC or vapor phase osmometry, and the values calculated from IR, UV or NMR absorptions with the assumption that the number of functions is equal to the number of molecules in the samples.

Ionic polymerization methods are well suited for the synthesis of macromonomers, because the active sites can exhibit — under well-chosen conditions — lifetimes that are long with respect to the duration of the polymerization. An additional advantage is that these methods yield linear polymers with polymerization degrees chosen at will, and with narrow molecular weight distributions.

One-step ionic synthesis of macromonomers can be performed in two different ways:

1.1 By ionic initiation

The polymerization of a monomer is initiated by means of an unsaturated initiator. It has to be checked that no side reaction will involve the double bond and that initiation occurs by addition to the monomer. If the initiation reaction is fast and quantitative the degree of polymerization is determined by the molar ratio (monomer)/(initiator) and each macromolecule formed should carry an end-standing double bond. This very straightforward method unfortunately applies to only a few cases. Some examples should be quoted:

The anionic ring-opening polymerization of oxirane, initiated by an unsaturated alcoholate (potassium is the most convenient counter-ion):

$$CH_2{=}\underset{CH_3}{\overset{}{C}}{-}\langle\bigcirc\rangle{-}CH_2OK + n\ \underset{O}{CH_2{-}CH_2} \xrightarrow{(+\ H^+)}$$

$$CH_2{=}\underset{CH_3}{\overset{}{C}}{-}\langle\bigcirc\rangle{-}CH_2O(CH_2{-}CH_2O)_nH$$

The cationic polymerization of oxolane (THF), that can be initiated either by an unsaturated oxocarbenium salt, or by an unsaturated benzylium salt, whereby

the poly-THF molecules formed are fitted with a polymerizable double bond at chain end:

$$CH_2{=}\underset{CH_3}{\overset{}{C}}{-}CO^+,\ SbF_6^- + n\ \underset{O}{\langle\bigcirc\rangle} \xrightarrow{(+\ CH_3O^-)}$$

$$CH_2{=}\underset{CH_3}{\overset{}{C}}{-}COO(CH_2)_4{-}(O(CH_2)_4)_{n-1}{-}OCH_3$$

$$\underset{CH_3}{\overset{CH_2}{\diagdown}}C{-}\langle\bigcirc\rangle{-}CH_2^+,\ SbF_6^- + n\ \underset{O}{\langle\bigcirc\rangle} \xrightarrow{(+\ CH_3O^-)}$$

$$\underset{CH_3}{\overset{CH_2}{\diagdown}}C{-}\langle\bigcirc\rangle{-}CH_2(O(CH_2)_4)_n{-}OCH_3$$

In this latter example the efficiency of the benzylium salt as initiator is even enhanced by the unsaturation in para position.

1.2 By induced ionic deactivation

Once the polymerization of the monomer has been completed, induced deactivation of the active sites can be used to fit the chain end with a double bond. The "living" anion has to be reacted with an unsaturated electrophile (or the "living" cationic site with an unsaturated nucleophile). Again it has to be checked that the unsaturation of the deactivator does not get involved in side reactions. To decrease the probability of such side reactions it is often necessary to decrease the nucleophilicity of the carbanions, by an intermediate addition of 1.1-diphenylethylene, or of oxirane, prior to the reaction with the functional deactivator:

$$\overset{PS}{\sim}CH_2{-}\underset{\emptyset}{\overset{}{CH^-}}K^+ \xrightarrow[2.\ ClCH_2\emptyset{-}CH{=}CH_2]{1.\ CH_2{=}\underset{\emptyset}{\overset{\emptyset}{C}}}$$

$$\sim CH_2{-}CH{-}CH_2{-}\underset{\emptyset}{\overset{\emptyset}{C}}{-}CH_2{-}\langle\bigcirc\rangle\overset{CH_2}{\underset{}{\overset{\|}{CH}}}$$

$$\text{\textasciitilde\textasciitilde}^{PS}CH_2-CH^-K^+ \xrightarrow[\text{2. } CH_2-CH=COCl]{1. CH_2-CH_2 \text{ (O)}}$$

$$\text{\textasciitilde\textasciitilde\textasciitilde}CH_2-CH-CH_2-CH_2-O-CO-CH=CH_2+KCl$$

with $\varnothing$ substituents.

This method of synthesis applies to far more systems than the first method:

Anionic deactivation was applied successfully to the synthesis of polystyrene, poly(alkylmethacrylate), poly(vinylpyridine), polydiene, poly(ethylene oxide) macromonomers fitted at chain end with either styrene type or methacrylic ester unsaturation [4–6]. All these species are well-defined and of low polydispersity. The unsaturations at chain end have been characterized quantitatively.

Cationic deactivation was also applied to the synthesis of macromonomers [6,7], especially those of poly-oxolane (poly-THF), and of poly(N-tert.butyl aziridine) [8], since these two systems can be considered as truly "living". Styrene or methacrylate endgroups can be introduced upon deactivation as shown in these examples:

The method described above can be extended to the synthesis of bifunctional macromonomers, carrying a polymerizable unsaturation at each chain end. A bifunctional ionic initiator is used to start the polymerization process, and once the monomer is converted into polymer, the active sites at both chain ends are reacted with an unsaturated deactivator to yield the macromonomer. This procedure was applied chiefly to "living" polystyrene, polydiene, poly(ethylene oxide) and polyoxolane.

The cationic polymerization of vinylic monomers involves transfer reactions which prevent the above methods from being applicable to macromonomer synthesis. However, the "Inifer" method developed by Kennedy [9] takes advantage of the transfer reactions themselves to obtain well-defined functions at chain end, and with high yields. If these functions are subsequently treated with an adequate unsaturated reagent, macromonomers result. This two-step process was applied successfully to the synthesis of polyisobutene macromonomers [10].

Free radical polymerizations are characterized by the very short life-time of active sites. To obtain functionalized chain ends, the best way is to add a functional transfer agent. These transfer reactions exhibit an additional advantage: they allow an approximate control of

$$\text{\textasciitilde\textasciitilde\textasciitilde}CH_2-CH_2-N-CH_2CH_2-\overset{\oplus}{N}\underset{CH_2}{\overset{CH_2}{\diagdown}} \quad \xrightarrow{+CH_2=\overset{CH_3}{\underset{}{C}}-COOH}$$
with $C(CH_3)_3$ and $C(CH_3)_3$ substituents

$$\text{\textasciitilde\textasciitilde\textasciitilde}CH_2-CH_2-N-CH_2-CH_2-N-CH_2-CH_2-OCO-C\overset{CH_2}{\underset{CH_3}{\diagup}}$$
with $C(CH_3)_3$ and $C(CH_3)_3$ substituents

$$\text{\textasciitilde}^{PTHF}O(CH_2)_4O(CH_2)_4^+O\!\!<\!\!\rceil, SbF_6^- \xrightarrow{+KO-CH_2-\varnothing-C\overset{CH_2}{\underset{CH_3}{\diagup}}}$$

$$\text{\textasciitilde\textasciitilde}O(CH_2)_4-O(CH_2)_4-O-CH_2-\langle\bigcirc\rangle-C\overset{CH_2}{\underset{CH_3}{\diagup}}$$

In the case of dioxolane, and other similar monomers, transfer reactions are known to occur, and this method cannot be applied, but other possible ways to synthesize macromonomers are presently investigated.

the molecular weight of the polymer formed. In order to attain very high functionalization yields it is advisable to use a free radical initiator that carries the same function as the transfer agent. Several two-step macromonomer syntheses based on the above principle have been developed by Yamashita [11]. The first step involves synthesis of a polymer chain fitted with alcohol or carboxylic functions at chain end:

$$\sim\sim CH_2-\underset{\underset{R}{|}}{CH}{}^{\cdot}+HS(CH_2)_2OH \longrightarrow \sim\sim CH_2-\underset{\underset{R}{|}}{CH}-S(CH_2)_2OH$$

$$\sim\sim CH_2-\underset{\underset{R}{|}}{CH}{}^{\cdot}+HS-CH_2-COOH \longrightarrow \sim\sim CH_2-\underset{\underset{R}{|}}{CH}-S-CH_2-COOH$$

Thereafter these functions are reacted with methacryloyl chloride or with glycidyl methacrylate, respectively, to yield the ω-methacryloyl macromonomers. This method has been applied to a large variety of monomers.

Polycondensation reactions are known to yield α, ω-difunctional macromolecules. The self-condensation of an unsymmetrical monomer AB is of special interest, since all macromolecules formed are fitted with an A function at one end and with a B function at the other. Upon reaction of one of them with an adequate unsaturated reagent macromonomers are formed. A

$$\sim\sim CH_2-COOH+CH_2-\underset{\underset{O}{\diagdown\diagup}}{CH}-CH_2-OCO\overset{\overset{CH_3}{|}}{C}=CH_2 \longrightarrow \sim\sim CH_2-COOCH_2-\underset{\underset{OH}{|}}{CH}-CH_2-O-CO-\overset{\overset{CH_3}{|}}{C}=CH_2$$

$$\sim\sim CH_2-OH+ClCO-\underset{\underset{CH_3}{|}}{C}=CH_2 \longrightarrow \sim\sim CH_2-O-CO-\underset{\underset{CH_3}{|}}{C}=CH_2$$

In a similar fashion free radical telomerization has been applied to the synthesis of macromonomers. Carbon tetrachloride is an efficient telogen which yields telomers fitted with $-CCl_3$ functions at chain end. This telomer can again serve as a telogen upon reaction with allyl alcohol, whereupon a 1:1 adduct is formed: the polymer chains are thus fitted with end-standing alcohol functions: to obtain the macromonomers these functions are reacted with methacryloyl chloride. This efficient method was developed by Pietrasanta [12] and was chiefly applied to the synthesis of short-chain polyvinyl chloride macromonomers and of fluorine containing macromonomers.

recent example is the synthesis of polyoxyphenylene macromonomers [13].

In this connection, the synthesis of polyamine macromonomers by Tsuruta [14] should be quoted: upon reaction of p-divinylbenzene with a secondary diamine a 1:1 adduct is formed in high yields.

$$CCl_4+nCH_2=\underset{\underset{Cl}{|}}{CH} \xrightarrow[\text{initiators}]{\text{Redox}} CCl_3-(CH_2-\underset{\underset{Cl}{|}}{CH})_n-Cl$$

$$CCl_3-(CH_2-CHCl)_n-Cl + CH_2=CH-CH_2OH \rightarrow HOCH_2-CHCl-CH_2-CCl_2-(CH_2-CHCl)_n-Cl$$

$$(I)$$

$$(I) + CH_2=CH-COCl \rightarrow CH_2=CH-COO-CH_2-CHCl-CH_2-CCl_2-(CH_2-CHCl)_n-Cl$$

$$CH_2=CH-\langle\bigcirc\rangle-CH=CH_2+HN(CH_2)_2-NH \longrightarrow CH_2=CH-\langle\bigcirc\rangle-CH_2CH_2-N(CH_2)_2-NH$$
$$\underset{CH_3}{|} \qquad \underset{CH_3}{|} \qquad\qquad\qquad\qquad\qquad\qquad \underset{CH_3}{|} \qquad \underset{CH_3}{|}$$

This adduct can undergo self condensation (initiated by weak bases). This reaction can also be considered as an anionic polymerization involving transfer after each growth step. In any case, all the polycondensate molecules exhibit a styrene unsaturation at one chain end and a secondary amine at the other.

$$CH_2=CH-\langle\bigcirc\rangle-CH_2-CH_2\Big[N(CH_2)_2-N-CH_2CH_2-\langle\bigcirc\rangle-CH_2-CH_2\Big]_n N(CH_2)_2-NH$$
$$\underset{CH_3}{|} \qquad \underset{CH_3}{|} \qquad\qquad\qquad\qquad\qquad \underset{CH_3}{|} \qquad \underset{CH_3}{|}$$

2. Homopolymerization of macromonomers

Owing to their relatively high molecular weights ($10^3 - 3 \times 10^4$) the molar concentration of macromonomers in a reaction medium is always low (say 0.01 to 0.5 mol $\cdot$ 1^{-1}). It can thus be expected that the polymerization degrees attained upon free radical homopolymerization of the macromonomer will remain low [15]. This was confirmed experimentally.

The anionic polymerization of macromonomers, under well-defined conditions, is possible, and even involves some control of the polymerization degrees [15].

In a polymacromonomer each unit of the main chain carries a graft. These macromolecules are very compact, and they exhibit a very high segment density within the coil. As a consequence, their hydrodynamic volume is very small as compared to that of a linear

homologue of same molecular weight. This was confirmed experimentally: their SEC elution volumes are larger, their limiting viscosity numbers and their radii of gyration are smaller, showing that the polymacromonomers are far more compact than the corresponding linear polymers (Table 1). Further measurements are presently in progress to study the influence of the length of the graft (i. e. that of the original macromonomer) on the hydrodynamic volume.

3. Synthesis of graft copolymers

The free radical copolymerization of a macromonomer with a vinylic or acrylic comonomer yields a graft copolymer, each macromonomer incorporated resulting in a graft. Such a copolymerization system is generally close to the non-azeotropic ideal case [15–17], owing to the low molar concentration of the macromonomer, already mentioned. This implies that the classical equation for the instantaneous copolymer composition reduces to

$$\frac{d[A]}{d[M]} = r_a \frac{[A]}{[M]}$$

$[A]$ and $d[A]$ refer to the comonomer and $[M]$ and $d[M]$ to the macromonomer. One single parameter, r_a, the reactivity ratio of the comonomer, has to be taken into consideration. This is advantageous, since the reactivity ratio of the macromonomer is not easily accessible.

The kinetics of copolymerization was studied in several laboratories. Size exclusion chromatography was used to determine, after given reaction times, the amount of copolymer formed and the amount of unconsumed macromonomer (Fig. 1). It was established that the rate of consumption of comonomer and of macromonomer both follow first order rate laws, as shown in Figure 2, up to conversions of the order of 60%. The ratio of the two slopes is equal to the

Table 1. Radii of gyration of polymacromonomers measured in benzene d_6

Ref.	Branch M_w (LS)	Polymacromonomer M_w (LS)	R_G (Å)	Linear equivalent R_G[b]	R_G[c]
H 3	1000	21000	26	40	–
H 3	1100	73000	45	82	–
2751	3000	97000	77	101	–
2751	3000[a]	97000	47	–	85

[a]) measured in cyclohexane d_{12} at 35 °C.
[b]) calculated from the relation $R_G = 7.5 \times 10^{-2}$ $M^{0.635}$ valid for linear polystyrene in benzene.
[c]) calculated from the relation $R_G = 0.27 \times M^{0.5}$ valid for linear polystyrene in θ conditions.

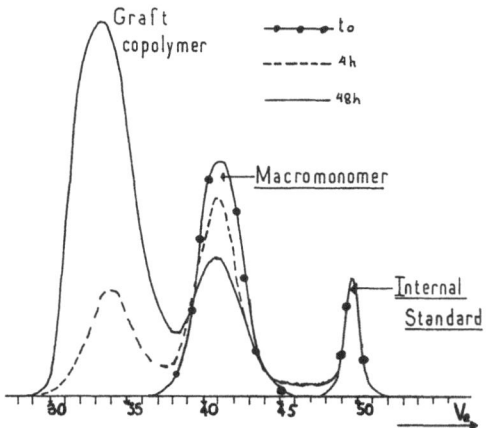

Fig. 1. Size exclusion chromatography (GPC) diagrams of the reaction mixture of a PEO macromonomer copolymerized with styrene at various times

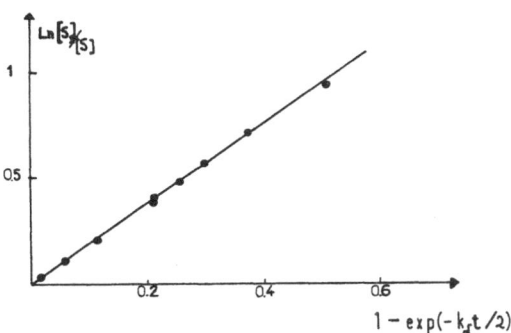

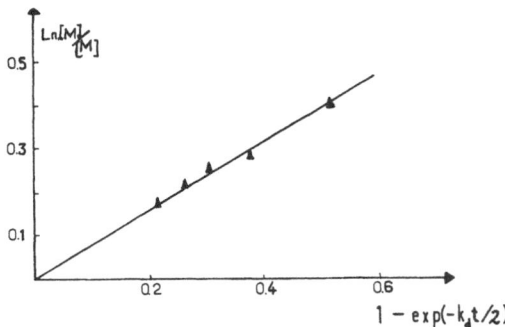

Fig. 2. First order rate plots of the consumption of styrene, S, and of the macromonomer, M, as a function of time

reactivity ratio r_a, thus confirming our assumption concerning ideality.

The instantaneous composition of the graft copolymer formed may differ from that of the mixture it originates from. Fluctuations in composition are therefore unavoidable within a sample obtained to finite conversions. The magnitude of these fluctuations is low, however, in most cases, and it is hardly influenced by the value of r_m.

Difficulties may be involved in the separation of the unconsumed macromonomer from the graft copolymer formed. It is well known that the behaviour of a graft copolymer in a solvent largely depends upon the affinity of the grafts for that solvent. There are cases in which the remaining macromonomer can be removed quantitatively, but in other cases emulsions are formed, preventing any separation.

The distribution of the grafts along the backbone chain should be random if account is taken of the kinetic results already mentioned. However no detailed investigation has yet been carried out. Fractionation of such samples yields copolymers of near constant composition over a wide range of molecular weights, thus confirming the randomness of the graft distribution along the backbone chain.

It thus appears that the graft copolymer synthesis by copolymerization of a macromonomer with a comonomer is easier and far more general that the classical "grafting on" and "grafting from" methods [18]. The species formed are rather well-defined, since grafts of known length are distributed at random along the chain.

The domain of application of graft copolymers steadily extends. The interest for these species arises from the fact that the interactions between chemically unlike sequences cause phase separation to occur. In solution aggregates and micelles are formed whereas in the bulk domain structures build up. Graft copolymers can act as compatibilizers (between the individual homopolymers), or as surface modifiers, or as emulsions producing agents. This latter application is specifically expected from graft copolymers constituted of a hydrophobic backbone carrying water soluble grafts.

The main interest of macromonomers is the easy access to graft copolymers. There are however a number of other potential applications of macromonomers, especially those arising from the reactivity of the end-standing double bond. Bifunctional macromonomers also exhibit specific properties which desserve investigations. They can be used in chain extension reactions, in block copolymer synthesis, and also quite obviously as crosslinking agents. They continue to receive increasing interest for due reasons.

References

1. Bamford CH, White EFT (1958) Trans Farad Soc 54:268
2. Greber G, Balciunas A (1963) Makromol Chem 69:193

3. Gillman KF, Senogles E (1967) Polym Letters 5:477
4. Schulz G, Milkovich R (1982) J Appl Polym Sci 27:4773; (1984) J Polym Sci, Polym Chem Ed 22:1633
5. Rempp P, Franta E (1984) Adv Polym Sci 58:1
6. Rempp P, Lutz P, Masson P, Franta E (1984) Makromol Chem Suppl 8:3
7. Asami R, Takaki M, Kita K, Asakura E (1980) Polym Bull 2:713
8. Goethals E, Vlegels M (1980) Polym Bull 4:521
9. Kennedy JP (1980) Polym J 12:609
10. Kennedy JP, Hiza M (1983) J Polym Sci, Polym Chem Ed 21:1033
11. Yamashita Y, Ito K, Mizuno H, Okada K (1982) Polym J 14:255
12. Boutevin B, Pietrasanta Y, Taha M, El Sarraf T (1983) Makromol Chem 184:2401
13. Percec V, Rinaldi PL, Auman BC (1983) Polym Bull 10:397
14. Nishimura T, Maeda M, Nitadori Y, Tsuruta T (1982) Makromol Chem 183:29
15. Rempp P, Lutz P, Masson P, Chaumont P, Franta E (1985) Makromol Chem Suppl 13:47
16. Kennedy JP, Lo CY (1982) Polym Bull 8:63
17. Revillon A, Hamaide T (1982) Polym Bull 6:235
18. Rempp R, Franta E, Herz J (1981) McGrath (ed) ACS Symp Series, Anionic Polym, p 59

Received November 22, 1985;
accepted December 19, 1985

Authors' address:

P. Rempp
Institut Charles Sadron
6, rue Boussingault
67083 Strasbourg Cedex, France

Synthesis and polymeranalogous reaction of poly(isocyanato alcanoic acid trialkylsilyl esters)*)

W. Mormann and R. Sikora

Universität-Gesamthochschule Siegen, Fachbereich 8, Siegen, F.R.G.

Abstract: Several isocyanato alcanoic acid trialkylsilyl esters were anionically polymerized to substituted 1-polyamides. Polymeranalogous reactions as hydrolysis and transesterification were carried out and show the applicability of the poly(isocyanates) as reactive stiff molecules.

Key words: Isocyanato alcanoic acid trialkylsilyl ester, polyamide, poly(isocyanate).

Introduction

Poly(isocyanates), also called 1-poly(amides), are a well known class of chain growth polymers with a rigid rodlike structure [1, 2]. Other polymers with stiff chains are poly(benzylglutamates) [3], poly(isonitriles) [4] and certain poly(saccharides) [5].

Rodlike macromolecules require theoretical and practical interest [6, 7] because of their behaviour in solution and their tendency to form lyotropic and (or) thermotropic liquid crystalline phases. Stiff molecules with reactive functional groups in the sidechain have not been described with few exceptions: poly(2.4-diisocyanatotoluene) and poly(3-(1-isocyanatoethyl)-phenyl isocyanate), which however have been neither isolated nor characterised [8].

Functionalized poly(isocanates) are also of interest because of their structure in solution as compared with that of other functionalized polymers such as acrylic acid derivatives.

As silylation is widely used for the reversible protection of active hydrogen bearing groups and, for example, silylation of phenolic hydroxy groups has been successfully applied to anionic polymerisation [9], isocyanato alcanoic acid trialkylsilyl esters were chosen as suitable monomers meeting the requirements of both polymerisability and sufficient reactivity of the functional group.

Results and discussion

Synthesis of the poly(isocyanates)

Monomeric isocyanato propanoic and butyric acid trialkylsilyl esters *1 a–1 c* were prepared from succinic and glutaric anhydride by reaction with azidotrimethyl or -tert. butyl-dimethylsilane [10]. Polymerisation was carried out in toluene using n-butyl lithium as initiator [11].

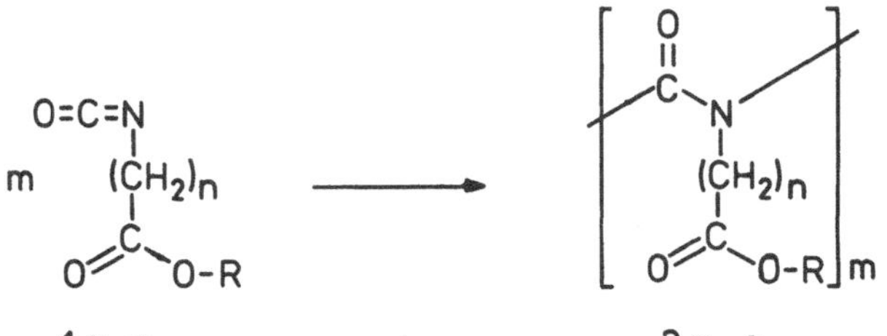

1 a: n=2; R = Si(CH₃)₃

   b: n=2; R = Si(CH₃)₂C(CH₃)₃

   c: n=3; R = Si(CH₃)₃

The poly(isocyanates) *2 a* and *2 c* are extremely sensitive to moisture and were handled in a glove box under

*) Lecture presented during the 32nd Annual Meeting of the Kolloid-Gesellschaft, Berlin October 2–4, 1985.

dry nitrogen. *2 b* with the tert.butyl-dimethylsilyl protecting group is much more stable to humidity and could be handled without severe precautions. Some of the properties of the poly(isocyanates) are summarized in Table 1.

The thermal behavior of the 1-poly(amides) was investigated by differential scanning calorimetry. Each of the three poly(isocyanatocarboxylic acid trialkylsilyl esters) showed at the beginning of the macroscopic melting temperature an exothermic signal followed by

Polymeranalogous reactions

Desilylation reactions

Hydrolysis of poly(isocyanatopropanoic acid trimethylsilyl ester) *2 a* was easily achieved in aqueous ethanolic suspension or solution. Stirring at room temperature for 24 hours yielded the solution of the polymeric isocyanatopropanoic acid *2 d* which could be isolated as a white powder, and had a pH of 3–4 in aqueous solution.

the endothermic melting peak. The exotherm is probably due to a depolymerisation-trimerisation reaction because on slow heating (Fig. 1) no melting endotherm is observed which is in agreement with the expected melting points of the trimers or monomers which normally are much lower than those of the 1-poly(amides).

Desilylation of the poly(isocyanatopropanoic acid tert.butyl-dimethylsilylester) *2 b* in aqueous alcohol required prolonged heating and gave only insoluble products having no acidic properties. Heating of *2 a* and *2 b* in water always resulted in evolution of carbon

Table 1. Properties of the poly(isocyanates)

Polymer	Softening point/°C	Melting point/°C	$[\eta]$/dl/g (solvent)
2 a	125	160 dec.	0.21 (toluene)
2 b	170	192	0.81 (toluene) 0.47 (CHCl₃)
2 c	—	150 dec.	0.20 (toluene)
2 d	—	154 dec.	see Figure 2

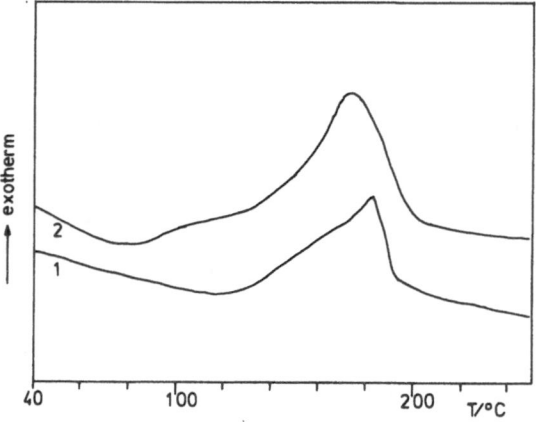

Fig. 1. DSC diagrams of *2 b* (1) and *2 d* in air, heating rate 10 K/min

dioxide. The same was observed when the polymeric acid *2 d* was heated above its melting point. Obviously the poly(isocyanatopropanoic acid) is depolymerized to monomer and trimer which can undergo polycondensation to mixed anhydrides which form 3-poly(amide) structures by loss of carbon dioxide. Spectroscopical, analytical and DSC data support this assumption.

Successful hydrolysis of *2 b* without significant degradation was possible in alcoholic suspension containing hydrogen fluoride at room temperature. The polymeric acids *2 d* showed similar properties and had the concentration dependence of the reduced viscosity which is typical of polyelectrolytes (Table 1 and Fig. 2).

(identified by comparison with authentic material [11]) but in addition the presence of carboxyl groups. Attempts to transform a poly(isocyanatopropanoic acid methyl ester) into the corresponding trimethylsilyl ester [12] were unsuccessful for the polymer as well as for the corresponding monomer, when they were reacted with iodotrimethylsilane.

Resilylation of *2 d* was possible with N.O-bis(trimethylsilyl)acetamide. Esterification of *2 d* with diazomethane and anhydride formation of *2 b* with acid fluorides were successful in preliminary experiments and are currently under investigation.

These experiments show that poly(isocyanatoalcanoic acid silylesters) are suitable polymers for polymeranalogous reactions.

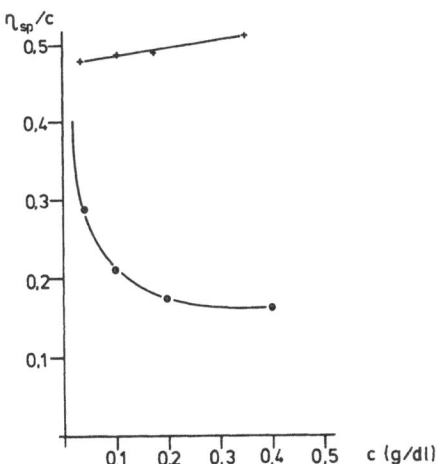

Transesterification

Transesterification of *2 a* or esterification of *2 d* were possible but not quantitative with methanol in boiling toluene. The infrared spectrum of the product showed the carbonyl stretching frequency of the ester group

References

1. Bur AJ, Fetters LJ (1976) Chem Rev 76:727
2. Tiger RP, Sarynina LI, Entelis SG (1972) Russ Chem Rev, Engl Transl 41(9):744
3. Flory PJ, Leonard Jr WJ (1965) J Amer Chem Soc 87:2102
4. Van der Eijk JM, Richters VEM, Nolte RJM, Drenth W (1984) Recl Trav Chim Pays-Bas 103:46-50
5. Tseng SL, Laivins GV, Gray DG (1982) Macromolecules 15:1262
6. Ballauf M, Macromolecules, to be published
7. Schmidt M, Paradossi G, Burchard W (1985) Macromol Chem Rapid Commun 6:767
8. Sashoua VE, Sweeny W, Tietz RF (1960) J Amer Chem Soc 82:886
9. Hirai A, Yamaguchi K, Takenaka K, Suzuki K, Nakahama S, Yamazaki N (1982) Macromol Chem Rapid Commun 3:941
10. Kricheldorf HR (1972) Chem Ber 105:3958
11. Mormann W, Schwabe A, Sikora R (1986) Makromol Chem 187, to be published
12. Jung ME, Lyster MA (1977) J Amer Chem Soc 99:968

Received December 3, 1985; accepted December 19, 1985

Authors' address:

W. Mormann
Universität-Gesamthochschule Siegen
Labor für Makromolekulare Chemie
Adolf-Reichwein-Str. 2
D-5900 Siegen 21, F.R.G.

Fig. 2. Reduced viscosity of *2 b* (+) and *2 b* and *2 d* (O) as a function of concentration (solvent for *2 b* chloroform, for *2 d* water/ethanol

Progress in Colloid & Polymer Science   Progr Colloid & Polymer Sci 72:122–128 (1986)

Hydrogels based on poly(2-hydroxyethyl methacrylate) with glucose moieties*)

G. Koßmehl, J. Volkheimer, and H. Schäfer

Institut für Organische Chemie der Freien Universität Berlin, Berlin, F.R.G., and
Titmus Eurocon Kontaktlinsen GmbH, Aschaffenburg, F.R.G.

Abstract: Hydrogels (crosslinked polymers with a definite water content) with glucose
moieties, prepared by radical copolymerization of 3-0-methacryloyl-1,2; 5,6-diisopro-
pylidene-α-D-glucofuranose, 2-hydroxyethyl methacrylate, the crosslinker ethylene
glycol dimethacrylate and acid hydrolysis of the isopropylidene protecting groups, were
examined for linear expansion to hydration, weight-percent hydration, water content on
a dry basis, refractive index, oxygen permeability and the percental transmission of visible light.

Key words: 3-0-methacryloyl-1,2;5,6-diisopropylidene-α-D-glucofuranose, contact
lenses, polymethacrylates with glucose moieties, hydrogels with glucose moieties, oxygen permeability, HEMA.

1. Introduction

In 1981, Tanaka et al. described for the first time (in a
patent [1]) the synthesis and the characterization of hydrogels (crosslinked polymers with a defined weight-
percent hydration) with glucose moieties as suitable
materials for soft contact lenses with high oxygen permeability and good compatability. Among other hydrogel samples, two hydrogel samples have also been
described, prepared from 3-0-methacryloyl-1,2;5,6-
diisopropylidene-α-D-glucofuranose (3-MDP-Glu),

*) Lecture presented during the 32nd Annual Meeting of the
Kolloid-Gesellschaft, Berlin October 2–4, 1985.

K 59

2-hydroxyethyl methacrylate (HEMA)

and ethylene glycol dimethacrylate (EGDM) as a
crosslinker

by radical copolymerization followed by acid hydrolysis for splitting off of the isopropylidene protecting
groups.

In our paper, 28 hydrogel samples of different composition, produced by radical copolymerization of 3-
MDP-Glu, HEMA and EGM with azobisisobutyronitrile (AIBN) as an initiator and acid hydrolysis, were
examined for linear expansion to hydration, weight-
percent hydration, water content on a dry basis, refractive index, oxygen permeability and percent trans-

mission of visible light as a function of the content of the starting monomers.

3-MDP-Glu was synthesized by a new method. We obtained 3-MDP-Glu as crystals free of inhibitor and in a higher yield than that described in the literature. The product was characterized spectroscopically, particularly by its ^{1}H-NMR-spectrum.

2. Experimental part

2.1 Preparation of 3-0-methacryloyl-1,2; 5,6-diisopropylidene-α-D-glucofuranose (3-MDP-Glu)

36.8 g (0.77 mol) of a dispersion (50%) of sodium hydride in mineral oil (Aldrich Chemie) were placed in a 1-l three-necked flask and washed twice with 50 ml ligroin (boiling range 100/110 °C). 500 ml of ligroin were added to the washed sodium hydride and then, while being stirred at 40 °C under nitrogen, 200.0 g (0.77 mol) 1,2; 5,6-diisopropylidene-α-D-glucofuranose [4] was added. The reaction mixture was stirred under nitrogen until the development of hydrogen ceased (approx. 5 h). Under nitrogen, 87.0 g (0.83 mol) methacryloyl chloride (methacryloyl chloride from Fluka, stabilized with 0.01% 2,6-ditert-butyl-p-cresole; freed from inhibitor by distillation) was added drop-wise (exothermic reaction) so that the temperature remained between 40 and 50 °C. After addition of the methacryloyl chloride, the mixture was stirred for one hour at 60 °C and filtered at room temperature. The ligroin phase was washed twice with 500 ml of a sodium hydroxide solution (5%), twice with 500 ml of water and dried over sodium sulfate. At a temperature below 30 °C and under reduced pressure, the ligroin was removed and 70 g of copper(I)-chloride (as an inhibitor) were added to the remaining syrup.

The following distillation (bp.: 103 °C; 0.09 hPa; bath temperature: 135 °C) gave 201 g (80% yield) of a colourless syrup. The distillation was carried out without a condensor; the collecting flask was cooled with ice-water and protected from daylight and neon light as 3-MDP-Glu, free of inhibitor, polymerizes quickly, even under light with a small amount of ultra-violet intensity. Immediately after the distillation, 100 ml of petroleum ether (boiling range: 40 °C/70 °C) was dissolved in the obtained syrup (201 g). The solution was left at a temperature of −10 °C for one day to crystallize. Afterwards, the −10 °C solvent was decanted from the 3-MDP-Glu crystals. The crystals were dried at room temperature. Yield: 190 g (= 75%); m. p.: 44–44.5 °C ([2]: 35 °C).

Crystalline 3-MDP-Glu does not polymerize even when exposed to daylight at 20 °C. If the crystals are melted, the melt does not recrystallize but polymerizes quickly.

($C_{16}H_{24}O_7$) (328.4) Calc. C 58.53 H 7.37
Found C 58.54 H 7.41
Ir (KBr): 1640 (m; C=C), 1730 (s; C=O).
MS (EI; 80 eV; 40 °C): m/e 313 (100% M−CH$_3$);
287 (6% M−C(CH$_3$)=CH$_2$); 255 (12% M−CH$_3$−CO−CH$_3$);
195 (9% M−CH$_3$−CO−CH$_3$−CH$_3$COOH);
101 (83% ┌O−C(CH$_3$)$_2$−O$^\oplus$=CH−CH$_2$−O┐).

^{1}H-NMR (270 MHz, DMSO$_{d6}$) δ (ppm): 6.10 (m; 1 H^i); 5.95 (d; J$_{a,b}$ = 3.8 Hz; 1 H^a); 5.75 (m; 1 H^h); 5.10 ("d"; J$_{c,d}$ = 3 Hz; 1 H^c); 4.60 ("d"; J$_{b,c}$ < 0.5 Hz; 1 H^b); 4.25 (m; J$_{e,f}$ = 6 Hz; 1 H^e); 4.20 (m; 1 H^d); 4.05 (double d; J$_{f,g}$ = 8.5 Hz; 1 H^f); 3.90 (double d; J$_{g,e}$ = 5 Hz; 1 H^g); 1.90 ("s"; 3 H^j); 1.25–1.45 (4s; 3 H^k; 3 H^l; 3 H^m; 3 H^n).

2.2 Preparation of the polymeric samples

Altogether, 28 polymeric samples with different concentrations of 3-MDP-Glu, HEMA and EGDM were produced (see Table 1). For weighing of HEMA, the EGDM portion in HEMA (0.14 mol%; determined by HPLC) was taken into consideration. The quantity from each weighing of 3-MDP-Glu and HEMA amounted to 100 mol%. Each run weighed 15 g (see Table 1). 7.5 mg of azobisisobutyronitrile (AIBN) was added to each mixture with the exception of nos. 7, 14, 21 and 28. The mixtures nos. 1–6, 8–13, 15–20 and

Table 1. Composition of the polymeric samples

No.	EGDM mol%	g	HEMA mol%	g	3-MDP-Glu mol%	g
1	0.2	0.01366	100	14.9863	0	0
2	0.2	0.01419	95	13.2324	5	1.7534
3	0.2	0.01463	90	11.7097	10	3.2757
4	0.2	0.01538	80	9.1963	20	5.7884
5	0.2	0.01654	65	6.1843	35	8.7993
6	0.2	0.01683	50	4.2594	50	10.7238
7	0.2	0.01809	0	0	100	14.9819
8	0.5	0.08163	100	14.9184	0	0
9	0.5	0.07736	95	13.1766	5	1.7460
10	0.5	0.07366	90	11.6635	10	3.2628
11	0.5	0.06755	80	9.1642	20	5.7682
12	0.5	0.06188	65	6.1656	35	8.7725
13	0.5	0.05553	50	4.2484	50	10.6961
14	0.5	0.04514	0	0	100	14.9549
15	1	0.19353	100	14.8065	0	0
16	1	0.18149	95	13.0847	5	1.7339
17	1	0.17102	90	11.5887	10	3.2415
18	1	0.15371	80	9.1114	20	5.7349
19	1	0.13684	65	6.1346	35	8.7285
20	1	0.11958	50	4.2302	50	10.6503
21	1	0.09001	0	0	100	14.9010
22	2	0.44250	100	14.5575	0	0
23	2	0.38544	95	12.9046	5	1.7100
24	2	0.36198	90	11.4382	10	3.1998
25	2	0.32308	80	9.0074	20	5.6695
26	2	0.28453	65	6.0737	35	8.6418
27	2	0.24606	50	4.1942	50	10.5598
28	2	0.17894	0	0	100	14.8211

22–27 were stirred for approximately 3 hours at room temperature until the 3-MDP-Glu crystals and the AIBN had dissolved in HEMA. In order to melt the 3-MDP-Glu crystals of the runs nos. 7, 14, 21 and 28 and to keep the melt liquid, these runs had to be stirred at a temperature of 50 °C. This resulted in polymerization before the AIBN had dissolved, making AIBN unnecessary. The mixtures were filled into polypropylene forms (capacity approx. 2 g, ø = 17 mm, h = 12 mm) and placed in a vacuum oven under nitrogen. The temperature was raised from 45 to 52 °C during a period of 3 h and left for 2 h at 52 °C. During the next 40 min the temperature was raised to 90 °C and left at this temperature for 2.5 h. The polypropylene forms were removed at room temperature. The polymeric samples were placed in a vacuum oven ($p < 1hPa$) to anneal over a period of 8 hours at 100 °C..

2.3 Hydrolysis of the polymeric samples

The copolymers were cut into discs (diameter: 11.0–12.1 mm; thickness: 0.19–0.34 mm), polished and according to a method described by Tanaka et al. [1], placed in 50 % formic acid for 0.5 h and then kept for 2 h in 6 N hydrochloric acid to ensure the splitting off of the isopropylidene protecting groups. Some of the discs which were only composed of 3-MDP-Glu and EGDM had to remain in the 6 N hydrochloric acid for a further 12 h in order to achieve a complete hydrolysis. After hydrolization, all the discs were put in a diluted solution of sodium carbonate (0.024 %) to neutralize and then kept for 10 days in a buffered isotonic sodium chloride solution of (1 l isotonic sodium chloride solution with 300 mosmol contained 3.04 g $Na_2HPO_4 \cdot 2 H_2O$, 0.84 g $NaH_2PO_4 \cdot H_2O$ and 8.00 g NaCl). During this time, the sodium chloride solution was exchanged three times.

2.4 Determination of weight-percent hydration, water content on a dry basis and linear expansion to hydration

The weighed, disc-shaped polymeric samples (dry weight) were kept in isotonic sodium chloride solution (which was exchanged three times) for 20 days, then weighed (wet weight). From this, the weight-percent hydration and the water content on a dry basis of the polymeric samples were calculated.

The wet weight of the hydrogel samples was specified from the hydrolized polymeric samples, prepared as described in 2.3. To achieve the dry weight of these hydrogel samples, they were placed in distilled water for 3 days to extract the salts (from the isotonic sodium chloride solution), then dried for 4 days at approximately 1 hPa over P_2O_5; firstly at 20 °C and afterwards at 40 °C until constant weight was acquired, then weighed (dry weight of the hydrogel samples).

To determine the linear expansion to hydration, the diameter of a polymeric sample was measured (diameter of a dry disc) and then the diameter of the sample after hydrolysis and storage in isotonic sodium chloride solution (diameter of the hydrogel sample).

2.5 Determination of the oxygen permeation

The oxygen permeation was measured at 27 °C with an Oxygen Flux Meter Type 920 from Schema Versatae, California, U.S.A., by the polarographic method of Fatt. The accuracy of the measurement amounted to ± 10 %.

3. Results and discussion

3.1 Structure of the polymers after hydrolysis

It is presumed that the isopropylidene protecting groups of the 3-MDP-Glu moieties in the polymeric samples were quantitatively split off. Consequently, the hydrogel samples have the following structural units:

3.2 Characterization of the polymers

3.2.1 Weight-percent hydration and water content on a dry basis

The water content of polymers is usually described by the "weight-percent hydration" and less frequently by the "water content on a dry basis".

The *weight-percent hydration* (in %) indicates the weight-percentage of water present in a polymer. It is determined by the weighing of samples in a dry state (dry weight) and in a swollen state (wet weight) using the following equation:

$$\text{weight-percent hydration [\%]} = \frac{\text{wet weight} - \text{dry weight}}{\text{wet weight}} \cdot 100$$

The *water content on a dry basis* (in %) on the other hand, gives information on how much weight percent

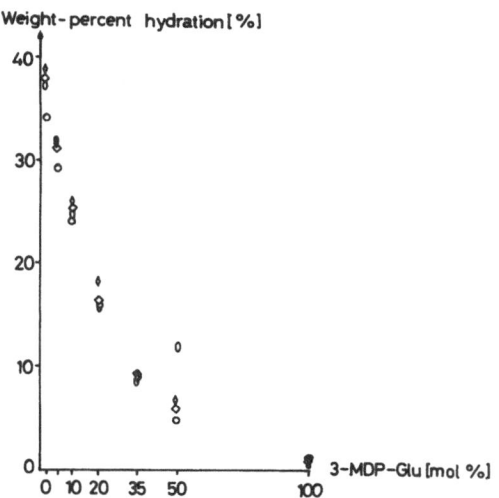

Fig. 1. Dependence of the weight-percent hydration on the composition of the polymeric samples made from 3-MDP-Glu, HEMA and EGDM after swelling in isotonic sodium chloride solution (pH = 7.2) at 20 °C. Concentration of EGDM: ◇: 0.2 mol %, ◇: 0.5 mol %, 0: 1.0 mol %, O: 2.0 mol %

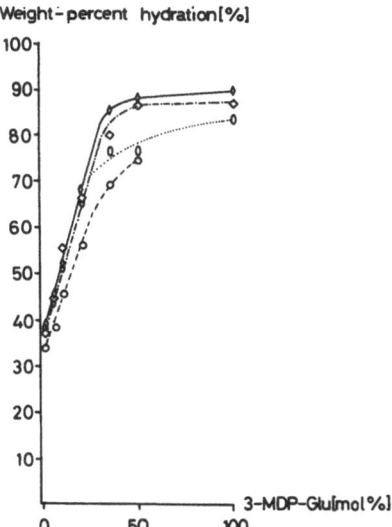

Fig. 2. Dependence of the weight-percent hydration on the composition of the polymeric samples made from 3-MDP-Glu, HEMA and EGDM after splitting off of the isopropylidene groups and swelling in isotonic sodium chloride solution (pH = 7.2) at 20 °C. Concentration of EGDM: ◇: 0.2 mol %, ◇: 0.5 mol %, 0: 1.0 mol %, O: 2.0 mol %

of water the dry polymer can absorb. This can be calculated by the following equation:

water content on a dry basis [%]

$$= \frac{\text{wet weight} - \text{dry weight}}{\text{dry weight}} \cdot 100$$

The polymeric samples which were produced by copolymerization, have decreasing values of weight-percent hydration with an increasing content of 3-MDP-Glu (due to the hydrophobic behaviour of this monomer, see Fig. 1).The different crosslinked polymeric samples from 3-MDP-Glu have a weight-percent hydration of approximately 1%. HEMA samples with different contents of the crosslinker EGDM show a weight-percent hydration of 35–39%. As expected, the weight-percent hydration of different crosslinked samples increases with decreasing amounts of the crosslinker.

The hydrophobic 3-MDP-Glu moieties of the polymeric samples were transformed into hydrophilic glucopyranose groups by acid hydrolysis. The weight-percent hydration of the hydrogel samples and the water content on a dry basis were determined as explained in the Experimental part.

As known from literature and also found by Koßmehl, Klaus and Schaefer [5] for the system HEMA with different crosslinkers in varying amounts, the weight-percent hydration of the hydrogels reduces with increasing content of a crosslinker (see Fig. 2). With an increasing content of glucose groups between 0 and 25 mol %, the weight-percent hydration of the disc-like hydrogel samples increases strongly in a linear way. The weight-percent hydration of the hydrogel samples containing 50 and 100 mol % glucose groups, reaches a value of saturation at approximately 90%.

The water content on a dry basis of the hydrogel samples gives results similar to those of the weight-percent hydration (see Fig. 3). Maximum values of up to 850% were determined for the water content on a dry basis. The largest rise was found in the region of 20 to 35 mol% of glucose moieties.

3.2.2 Linear expansion to hydration

The linear expansion to hydration describes the expansion of a polymer after complete swelling. The

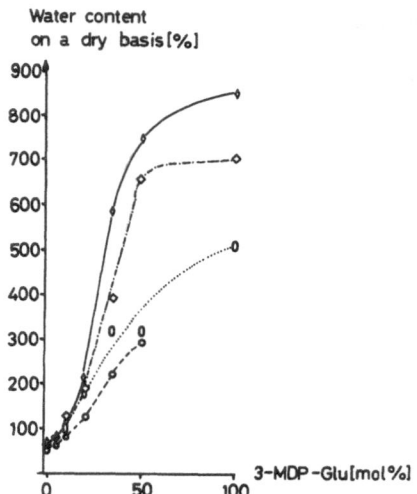

Fig. 3. Dependence of the water content on a dry basis on the composition of the polymeric samples made from 3-MDP-Glu, HEMA and EGDM after splitting off of the isopropylidene groups and swelling in isotonic sodium chloride solution (pH = 7.2) at 20 °C. Concentration of EGDM: ◊: 0.2 mol %, ◇: 0.5 mol %, 0: 1.0 mol %, O: 2.0 mol %

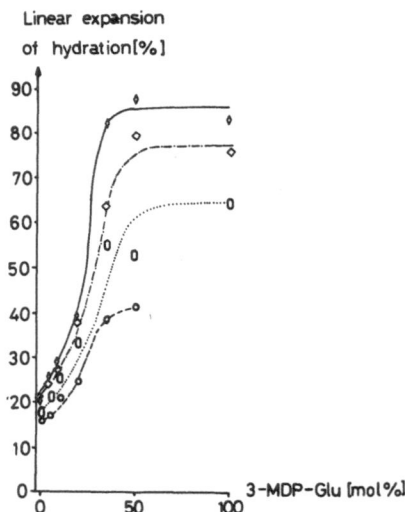

Fig. 4. Dependence of the linear expansion to hydration on the composition of the polymeric samples made from 3-MDP-Glu, HEMA and EGDM after splitting off of the isopropylidene groups and swelling in isotonic sodium chloride solution (pH = 7.2) at 20 °C. Concentration of EGDM: ◊: 0.2 mol %, ◇: 0.5 mol %, 0: 1.0 mol %, O: 2.0 mol %

linear expansion to hydration is calculated by the following equation:

$$\text{linear expansion to hydration [\%]}$$
$$= \frac{\text{swollen dimension} - \text{dry dimension}}{\text{dry dimension}} \cdot 100$$

In our case, 'dry dimension' means the diameter of the dry samples before hydrolysis. Figure 4 shows the relation between the linear expansion to hydration (in %) and the amount of glucose moieties (in mol %) in the hydrogel samples. The extreme increase of linear expansion to hydration lies in the region of between 20 and 35 mol % of glucose moieties. On the other hand, from a 50 % portion of glucose moieties onwards, the linear expansion to hydration hardly increases and reaches (for hydrogel samples crosslinked with 0.2 mol % EGDM) a saturation value of 85 %: The cohesive forces which hold the polymer network together, only allow a very slight increase of hydration in this region.

3.2.3 Refractive index

As Figure 5 shows, the dependence of the refractive index (n_D^{20}) of the hydrogel samples (see 2.4) on their weight-percent hydration is linear. This was also

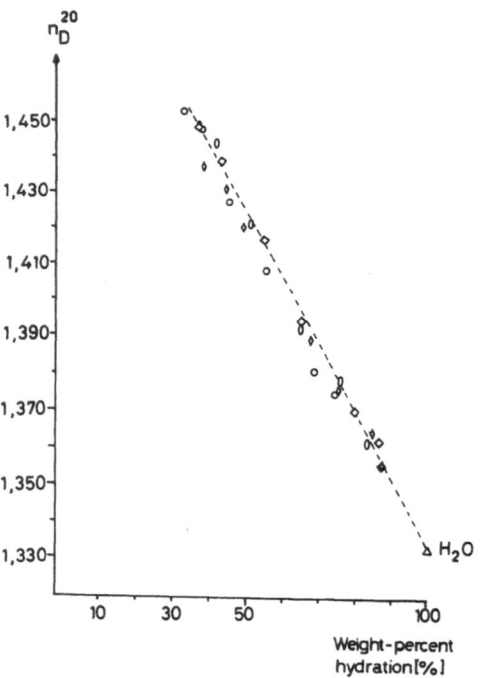

Fig. 5. Dependence of the refractive index of the hydrogel samples on their weight-percent hydration after swelling in isotonic sodium chloride solution (pH = 7.2) at 20 °C. Concentration of EGDM: ◊: 0.2 mol %, ◇: 0.5 mol %, 0: 1.0 mol %, O: 2.0 mol %

found by Klaus [6] for similarly structured hydrogels. With this linear dependency, it is possible to determine the weight-percent hydration by measuring the refractive index.

3.2.4 Oxygen permeability

Because of their weight-percent hydration, hydrogels have a certain permeability for gases. The gas permeability of a hydrogel is described by the so-called *permeability coefficient*. This is a strongly temperature-dependent [7] constant of materials [8] and therefore independent of the thickness of the hydrogel disc.

The permeability coefficient P for oxygen can be calculated by the following equation:

$$P = D \cdot K \left[\frac{\text{ml}(O_2) \cdot \text{cm}^2}{\text{ml} \cdot \text{s} \cdot \text{mm Hg}} \right].$$

D represents the diffusion constant (from the 1st law of Fick) and K the dissolving constant for oxygen (from the law of Henry). According to the given equation, the permeability coefficient describes the quantity of oxygen which passes through a given area (the hydrogel disc) per second, dependent on the difference of the partial oxygen pressures on both sides of the hydrogel disc.

In addition to the permeability coefficient, the oxygen permeability for hydrogels can be described by the *transmissivity* T, which is dependent on the thickness of the hydrogel disc. T is calculated by the equation:

$$T = \frac{P}{d} \left[\frac{\text{ml}(O_2)}{\text{cm}^2 \cdot \text{s} \cdot \text{mm Hg}} \right].$$

P represent the permeability coefficient and d is the thickness of the hydrogel disc. The transmissivity is the quantity of oxygen (in ml) which flows through a hydrogel area per second, having a partial oxygen pressure difference of 1 mm Hg between both sides of the hydrogel disc.

The *oxygen flux J* shows the quantity of oxygen which flows through a hydrogel disc of given thickness per unit of time. This can be calculated by the following equation:

$$J = \frac{P}{d} \cdot \Delta p \left[\frac{\mu\text{l}(O_2)}{\text{cm}^2 \cdot \text{h}} \right].$$

P is the permeability coefficient, d the thickness of the disc and Δp the difference of the partial oxygen pressures on both sides of the hydrogel disc.

Table 2. Oxygen permeation of the hydrogel samples nos. 1–25. Hydrogel samples nos. 7, 13, 20, 21, 26, 27, 28 had insufficient mechanical stability for measuring the oxygen permeation

No.	Disc thickness [mm]	P 10⁻¹¹	J	T · 10⁻⁹
1	0.32	8.7	11.5	2.7
2				
3	0.27	11.5	2.4	4.3
4	0.37	32.5	4.9	8.8
5	0.33	36.8	6.2	11.2
6	0.36	36.7	5.7	10.4
7				
8	0.26	7.2	1.7	2.8
9	0.26	9.3	2.0	3.6
10	0.35	24.9	4.0	7.1
11	0.33	26.4	4.5	8.1
12	0.27	30.1	6.2	11.2
13	0.27			
14	0.23	37.1	9.0	16.0
15	0.39	7.9	1.1	2.0
16	0.27	10.3	2.1	3.8
17	0.34	12.9	2.1	3.8
18	0.30	29.8	5.5	9.9
19	0.41	39.9	5.4	9.8
20				
21				
22	0.28	5.7	1.1	2.0
23	0.34	8.9	1.5	2.6
24	0.23	16.3	4.0	7.0
25	0.30	24.1	4.5	8.0

Table 2 presents the values for P, T and J of the examined hydrogel discs.

The measurements carried out show that the hydrogels produced from 3-MDP-Glu, HEMA and EGDM have oxygen permeability coefficients between 5 and 40 at 27 °C (see Fig. 6).

A linear connection exists between the weight-percent hydration of the hydrogels with values from 30 up to 70 % and the logarithm of P, as found by Tighe [7] for other hydrogels. As Figure 6 shows, no further increase of the oxygen permeability coefficient can be achieved, when the weight-percent hydration is raised above 70 %

3.2.5 Percent transmission of visible light

The hydrogel discs were placed between quartz plates and the transmission of visible light was measured in a wave-length region of 350 to 700 nm. The transmission of visible light of the examined hydrogel

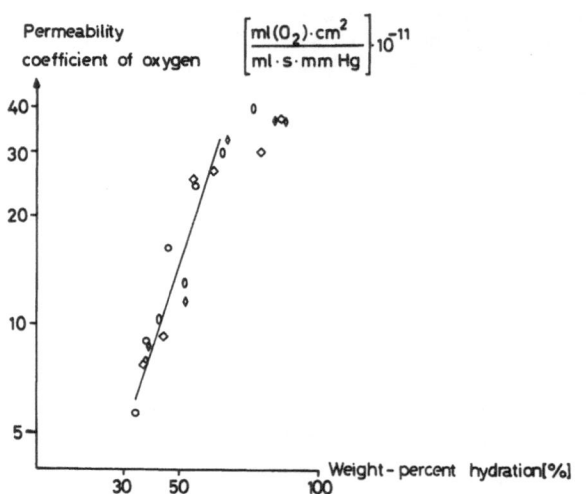

Fig. 6. Dependence of the permeability coefficient of oxygen of the hydrogel samples on their weight-percent hydration after swelling in isotermic sodium chloride solution (pH = 7.2) at 20 °C. Concentration of EGDM: ◊: 0.2 mol %, ◇: 0.5 mol %, 0: 1.0 mol %, O: 2.0 mol %

discs (for thickness, see Table 2) is approximately 90 %. For example, sample no. 11 has 91 % transmission at λ = 350 nm and 97 % transmission at λ = 700 nm.

References

1. 3200479 A1 (1982), Toyo Contact Lens Co, Inv: Tanaka K, Nagova A, Kanome S, Kuwana M, Nakajima T, Kazuhiko N, Nobuyuki T, Nagoya A, CA (1983) 98:149619q; JP · 57/116318 A2, Toyo Contact Lens Co, Inv: Tanaka K, Kanome S, Nakajima T, Nakada K, Toyoshima N, CA (1983) 98:149623 m
2. Black WAP, Dewar ET, Rutherford D (1963) J Chem Soc 4:4433
3. Kimura S, Imoto M (1961) Makromol Chem 50:155
4. Glen WL, Myers GS, Grant GA (1951) J Chem Soc 3:2568
5. Koßmehl G, Klaus N, Schäfer H (1984) Angew Makromol Chem 123/124:241
6. Klaus N (1986) Thesis, FU Berlin
7. Tighe BJ (1976) The British Polymer J 9:71
8. Kreiner C (1980) Kontaktlinsenchemie, Median-Verlag, Heidelberg

Received January 13, 1986;
accepted January 21, 1986

Authors' addresses:

Prof. Dr. G. Koßmehl and Dipl.-Chem. J. Volkheimer
Institut für Organische Chemie
der Freien Universität Berlin
Takustr. 3
D-1000 Berlin 33, F.R.G.

Dr. H. Schäfer
Titmus Eurocon Kontaktlinsen GmbH
Goldbacher Str. 57
D-8750 Aschaffenburg, F.R.G.

Progress in Colloid & Polymer Science   Progr Colloid & Polymer Sci 72:129–133 (1986)

Mixture analysis of polycondensates by mass spectrometry: macrocyclic compounds via Wittig polycondensation reactions*)**)

G. Holzmann, T. Kobilke, and G. Koßmehl

Institut für Organische Chemie, Freie Universität Berlin, Berlin, F.R.G.

Abstract: Mass spectrometrical methods reveal that reaction mixtures of Wittig polycondensation of bivalent ylides and bis-(triflouroacetyl)-benzene (or -thiophene) preferentially consist of cyclic compounds. The determination of ring units ($n = 2$–13) was performed by use of direct evaporation of solid samples, low energetical ionization and exact mass measurements, to determine elemental compositions. The structures of molecular ions present in the complex reaction mixtures were elucidated by metastable decompositions which show the characteristic fragmentation pattern of cyclic arylenes or heteroarylenes. General procedures of mass spectrometrical analysis are given. The influence of trifluoromethyl moieties on preferential formation of cyclic polycondensates is discussed.

Key words: Polycondensation via Wittig reaction, cyclic poly-vinylene -arylenes (-heteroarylenes), mixture analysis by mass spectrometry.

Introduction

Mass spectrometrical analysis of polycondensates (or polymers) is normally restricted by the involatility of the material or the formation of artifacts by thermal degradation or the ionizing methods used. Thus, the information obtained from mass spectra of complex mixtures of molecular and fragment ions is reduced. Therefore, we have used the registration of metastable decompositions [1] of selected ions in the reaction mixtures to characterize molecular sizes, terminal functions of linear products and structural details of oligomers. In continuation of our work [2] were report of general procedures of mass spectrometrical analysis of complex reaction mixtures obtained by

Wittig polycondensation reactions [3], generalized in Scheme 1.

Experimental

Polycondensation of bivalent ylides (Scheme 1) and bis(trifluoroacetyl)-benzene and/or thiophene by Wittig polycondensation reaction [4] was used yielding 80 to 90% of cyclic products and only 10% of linear material as by-products. The favourable formation of cyclic compounds is in contrast to the attempts [5] to prepare large ring systems by condensation of ylides and carbonyl precursors.

General procedures of synthetic routes are given in the literature [2]: the reaction mixtures thus obtained were cooled to room temperature, the solution filtered off from solid material and purified by LC (silicagel, Woelm TSC/CHCl₃) yielding fractions of cyclic components controlled by DC and HPLC. To the solution of cyclic compounds methanol was added, the solid material thus obtained was filtered off, washed with methanol to give either pure cyclic components of *4 a–d* (Scheme 1) or mixtures of purified reaction products (Table 1, *4 a–d*). These reaction products as mixtures or isolated distinct cyclic compounds were further analysed by mass spectrometrical methods, summarized in Scheme 2.

The determination of ring sizes of cyclic compounds and the elemental composition were performed by evaporation of the reaction

---

*) Part III of "MS-Studies of polymers"; Part I and II: Refs. [1, 2].
**) Dedicated to Professor Dr. Georg Manecke on his 70th birthday. Lecture presented during the 32nd Annual Meeting of the Kolloid-Gesellschaft, Berlin October 2–4, 1985.

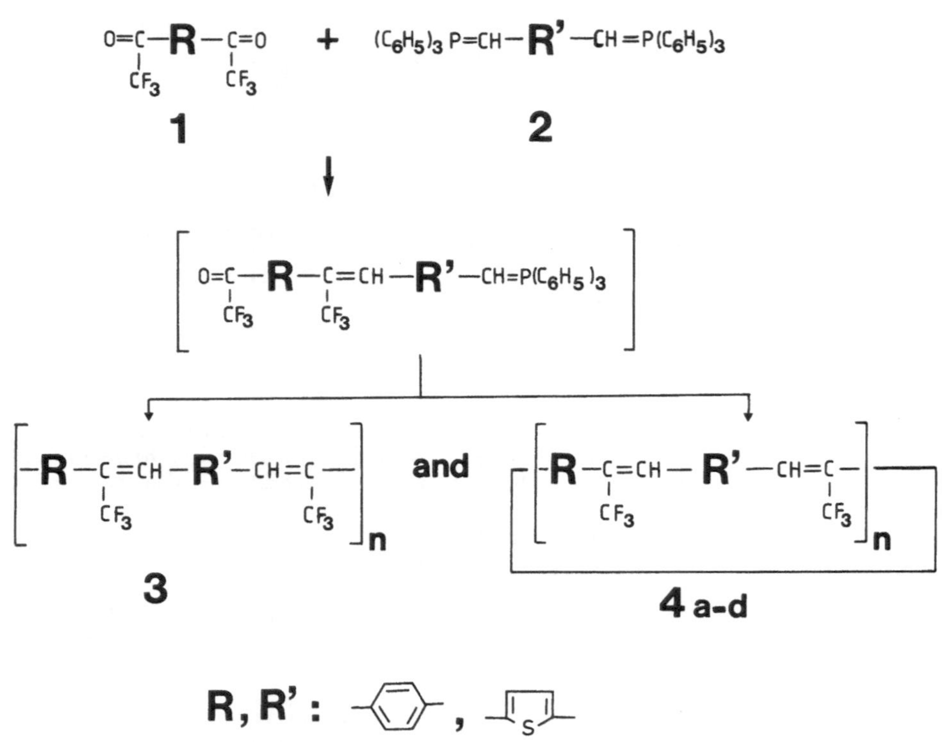

Scheme 1. Synthetic route to cyclic polycondensation products via Wittig polycondensation reaction : *4 a–d*

Table 1. Cyclic polycondensation products *4 a–d*: Distribution of cyclic compounds present in the EI spectra (80 eV) with ring units: *n*; molecular weights and elemental composition of typical examples are given

No.	Units of cyclic polycondensates: Sequences n	Units n	MZ of cycles (n=2)	Elemental Composition of cyclic dimers
4a	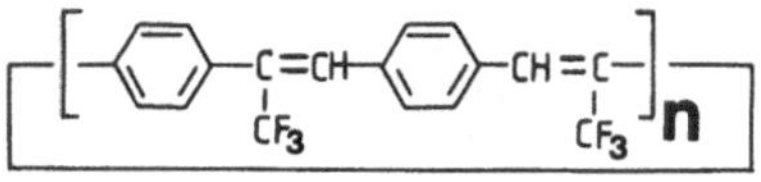	2-9	**680**	$C_{36}H_{20}F_{12}$
4b	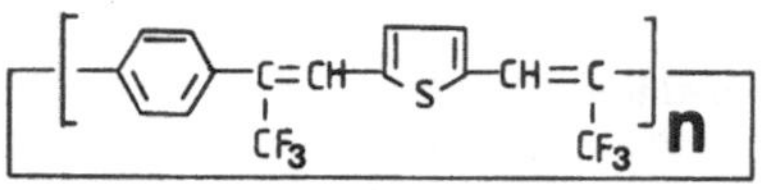	2-6	692	$C_{32}H_{16}F_{12}S_2$
4c	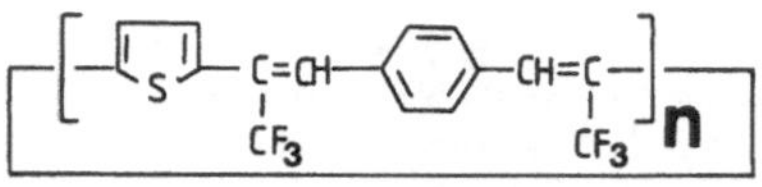	2-6	692	$C_{32}H_{16}F_{12}S_2$
4d	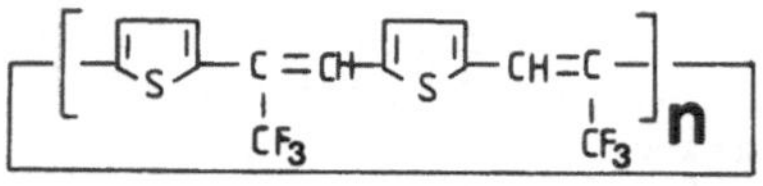	2-13	704	$C_{28}H_{12}F_{12}S_4$

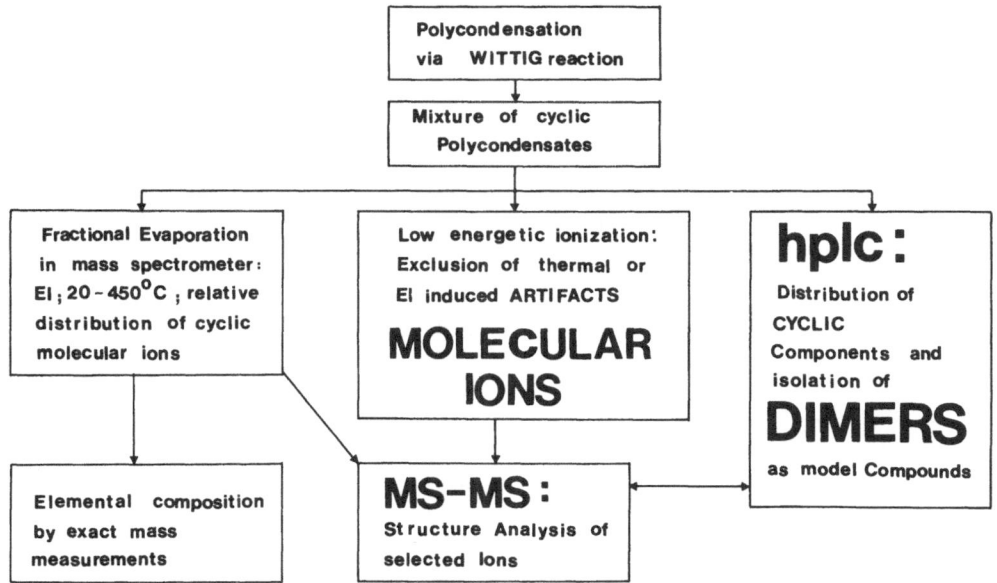

Scheme 2. Mixture analysis by mass spectrometrical methods of polycondenstation products

products and low energetical ionizing to exclude the detection of artifacts by thermal degradations. The solid material was evaporated in the ion source of a double focusing high resolving mass spectrometer (MAT 711, Varian MAT) and ionized by electron impact. Depending on the improved volatility of the material effected by trifluoromethyl groups cyclic compounds in the mass range up to m/z 3100 (Table 1) between 100–450 °C were detected. The mass chromatograms below the temperature of pyrolytical degradation enable the determination of highest and constant amount of cyclic components. At these temperatures the metastable decompositions of the molecular ions [1] were recorded to give specific fragmentation pattern of cyclic species used as fingerprint for the detection in the mixtures. These pattern were verified by study of model systems isolated by HPLC (Table 1: e. g. *4a*: m/z: 680).

Results and discussion

Mass spectrometrical mixture analysis reveals that in the polycondensation via Wittig reaction (Scheme 1) cyclic compounds are favourable formed (Table 1). The ring sizes observed in the electron impact mass spectra between 100–450 °C, the elemental composition of characteristic sequences and the molecular weight of typical cyclic compounds present in the reaction mixtures of polycondensates *4a–d* are summarized in Table 1. The isolated cyclic compounds were used as model systems for the detection of cyclic components in the reaction mixtures, e. g. m/z 680 is

the dimeric compound in the polycondensation mixture *4a*. This cyclic product was used to show the principles of the mass spectrometrical mixture analysis performed in this study.

The electron impact mass spectra of the polycondensation products *4a* at typical evaporation temperatures are depicted in Figure 1: different amounts of preferentially formed cyclic components given as molecular ions are detected. The distribution of ring sizes ($n = 2$–9) summarized as total ion currents is comparable to the relative intensities of HPLC analysis, so that fractional evapoartion of stable polycondensates may be used as qualitative information of molecular weight distribution in lower mass regions. In our case the detection of cyclic components is obviously limited by the mass range of the instrument (m/z 3600) and reflecting the extreme stabilities of the materials.

As an example of the favourable cyclic structures the metastable decomposition [1] of the dimer m/z: 680 of *4a* is shown in Figure 2. Besides typical abstraction of methyl radicals by rearrangement processes of the stilbene moiety loss of fluoro and trifluoromethyl radicals are observed.

Combination of these favourable fragmentations yields fragment ions which represent distinct intensity distributions used as finger-print for the analysis of cyclic components of analogous structures (Fig. 2).

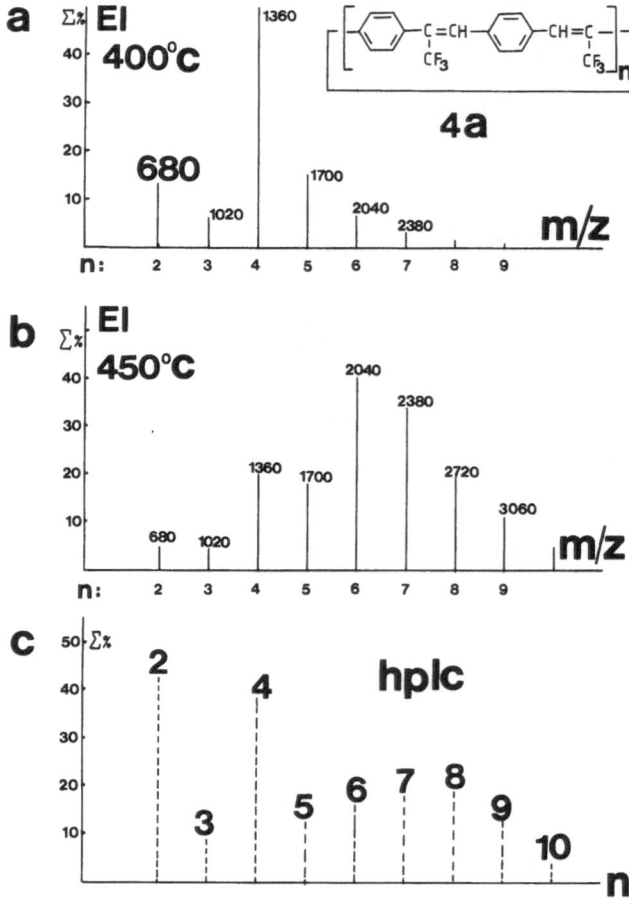

Fig. 1. Mixture analysis of polycondensation product *4a*; Electron impact mass spectra at 400 °C (a) and at 450 °C (b) compared to the distribution of components by HPLC (c); the repeating ring units (*n*) are given

Additional loss of SH-radicals characterizes the thiophene moieties in polycondensates *4b–d* by destruction of the thiophene rings. These compounds show enhanced ejection of HF as an assignment of interaction of the vinylene hydrogen atom and the adjacent trifluoromethyl function.

These characteristic fragmentations and the remarkable tendency of formation of cyclic polycondensates compared to nearly 10–20 % of linear products are explained by the influence of the trifluoromethyl function:

We assume that in the initial reaction step of the polycondensation reaction (Scheme 1) of bivalent trifluoroacetyl compounds *1* and ylides *2* intermediates of betain structures with preferential *E*-conformation [6] may be formed which in a self-condensation step or by dimerization favour the selective cyclizations to the cyclic compounds *4a–d*. The oligomerization and polycondensation to linear products *3* are surpressed (Scheme 1). The preferential *E*-conformation is explained by *H–F*-bonding of the adjacent fluoro and hydrogen atoms of the trifluoromethyl-vinylene units [6].

Conclusion

Mass spectrometrical analysis of compounds obtained by polycondensation reactions (Wittig) reveals that preferentially cyclic products are formed, achieved by trifluoromethyl functions of vinylene units of favourable *E*-conformation. The structures of

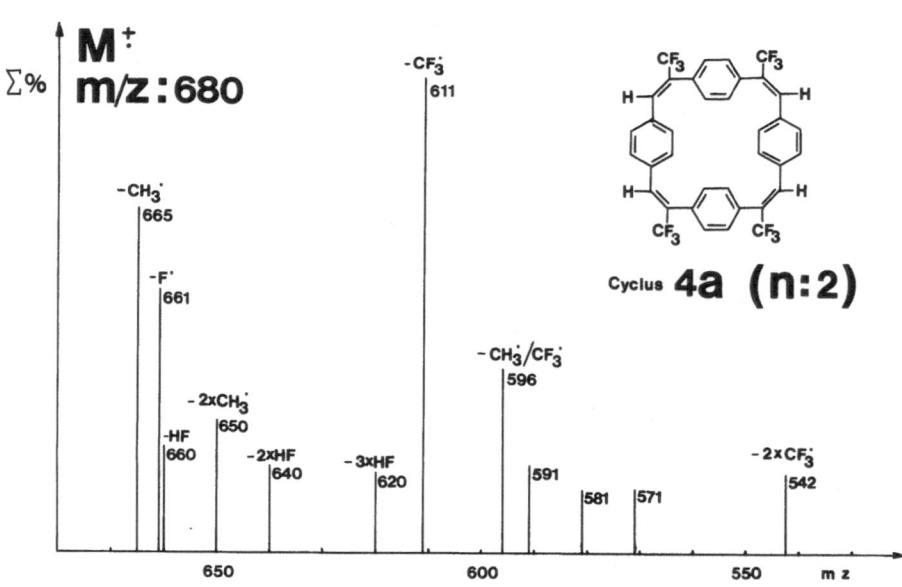

Fig. 2. Metastable decomposition of dimer present in the polycondensation product *4a* ($n = 2$; m/z: 680). The intensities are relative to the sum of all fragment ions ($\sum$ %). Main fragmentations are assigned; the elemental compositions of these ions are verified by exact mass measurements

distinct cyclic components were determined by metastable decompositions and the fragmentation pattern obtained were used as finger-print for analysis of cyclic species present in the reaction mixtures. The generalization of this mixture analysis to determine structures, repeating units, termal functions and molecular sizes of polycondensation reaction material is stressed.

Acknowledgements

The authors wish to thank Mrs. U. Ostwald for operating the mass spectrometer and Dr. W. Lamer for HPLC experiments.

References

1. Holzmann G, Koßmehl G (1980) Org Mass Spectrom 15:336
2. Holzmann G, Koßmehl G, Nuck R (1982) Makromol Chem 183:1711
3. Koßmehl G, Nuck R (1979) Chem Ber 112:2342
4. Kobilke T (1985) Ph D Thesis, Freie Universität Berlin
5. Thulin B, Wennerström O, Somfai I, Chmielarz B (1977) Acta Chem Scand B 31:135
6. Ruban G, Zobel D, Koßmehl G, Nuck R (1980) Chem Ber 113:3384

Received December 3, 1985;
accepted February 19, 1986

Authors' address:

Dr. Gerhard Holzmann
Institut für Organische Chemie
Freie Universität Berlin
Takustr. 3
D-1000 Berlin 33, F.R.G.

Subject Index